RAND NATIONAL SECURITY RESEARCH DIVISION

T0108946

The Economic Consequences of Investing in Shipbuilding

Case Studies in the United States and Sweden

Edward G. Keating, Irina Danescu, Dan Jenkins, James Black,
Robert Murphy, Deborah Peetz, Sarah H. Bana

For more information on this publication, visit www.rand.org/t/RR1036

Library of Congress Cataloging-in-Publication Data is available for this publication.

ISBN: 978-0-8330-9036-2

Published by the RAND Corporation, Santa Monica, Calif.

© Copyright 2015 RAND Corporation

RAND® is a registered trademark.

Cover: Littoral Combat Ship 6 (Jackson) *and 8* (Montgomery) *under construction in the Mobile River at Austal USA's site in Mobile, Alabama (photo by Irina Danescu).*

Support RAND

Make a tax-deductible charitable contribution at
www.rand.org/giving/contribute

www.rand.org

Preface

In 2014, the Australian Department of Defence engaged RAND to undertake a series of materiel studies and analysis activities.

This report on the economic consequences of shipbuilding is part of a larger RAND project for the Australian government entitled Analysis of Australian Shipbuilding Industry and Capabilities. The larger project is to inform Australian policymakers of the economics and feasibility of various strategies for the Australian shipbuilding industrial bases that produce or repair naval surface vessels.

One task under this broader project is entitled Economic Considerations. This task is to assess the relationship between Australia's maritime spending and the nation's levels of output, employment, and earnings. This report presents what the RAND research team has learned.

Questions regarding RAND Australia should be directed to Jennifer Moroney, at moroney@rand.org or 61 2 6243 4869.

This research was conducted within the Acquisition and Technology Policy Center of the RAND National Security Research Division (NSRD). NSRD conducts research and analysis on defense and national security topics for the U.S. and allied defense, foreign policy, homeland security, and intelligence communities and foundations and other nongovernmental organizations that support defense and national security analysis.

For more information on the Acquisition and Technology Policy Center, see http://www.rand.org/nsrd/ndri/centers/atp.html or contact the director (contact information is provided on the web page).

Contents

Figures and Tables

Figures

Tables

Summary

As part of a larger RAND study on the Australian shipbuilding production and repair industrial bases, this report discusses the economic consequences of shipbuilding in Australia.

This report is built around two different prospective paths for the Australian government (as well as a prospective hybridization of the paths). One path would be for the Australian government to pay Australia-based shipbuilders to build all Australian naval vessels. The opposite path would be for the Australian government to acquire ships for the Australian Navy from foreign providers.

We categorize the prospective economic consequences of these paths under two rubrics:

1. **Opportunity cost/displacement:** *What would individuals employed in Australian shipbuilding be doing otherwise?* If ships were purchased from foreign providers, those individuals who would have worked in shipbuilding would have different labor market outcomes, ranging from possible unemployment to working in a different type of manufacturing setting in Australia.
2. **Spin-offs/spillovers:** *To what extent would shipbuilding in Australia generate favorable spin-offs and spillovers?* We use the term *spin-off* to refer to a new firm that spins off an established firm. A *spillover* is a more general expression for a positive, perhaps unanticipated, consequence of a project, such as the firm being able to enter a different industry or trained workers moving to other employers.

To address these questions, this report uses analogies from the United States and Sweden to draw insights about the economic consequences of shipbuilding in Australia.

To inform and prepare for our case studies, we review the extensive literature on economic multipliers.

Economic Multipliers and Their Implications

While there is a sizeable literature on economic multipliers, its implications for the economic consequences of shipbuilding in Australia are uncertain.

The basic logic of an economic multiplier is straightforward. Suppose that the government spends $100 buying a good or service from a shipbuilder. The shipbuilder might then be expected to spend at least a portion of the $100 on inputs, such as labor or materials. The original $100 creates a cascade (i.e., multiples) of spending through the economy. That $100 spent on a shipbuilder results in additional spending by shipbuilder workers at local restaurants, which then hire additional workers who rent additional housing, and so forth.

Several studies have estimated economic multipliers associated with defense spending. Most of the resulting estimates are in the range of 1.7–1.9—i.e., $100 spent on a shipbuilder ultimately results in $170–190 worth of additional economic activity in the shipbuilder's region (inclusive of the original $100).

Economic multipliers may be lower (i.e., less than 1) if the increased spending displaces other economic activity. Studies looking at the Second World War often find multipliers less than 1 because increased defense spending displaced private sector spending. Deloitte Access Economics (2014) used a computable general equilibrium approach that, because of a full-employment assumption, resulted in an economic multiplier estimate near zero for the *Collins* program. On the other hand, if the spending at the shipbuilder results in favorable spillover effects into the economy, e.g., spin-offs into other industries, one could find an economic multiplier greater than the 1.7–1.9 range.

The near-zero multiplier from Deloitte Access Economics (2014) presents the argument that shipbuilding has no impact. The contrary argument is that shipbuilding has large-scale beneficial effects (i.e., a large multiplier), espoused by Roos (2014) and Economic Development Board South Australia (2014).

The Australian government faces a trade-off between potentially higher costs of indigenous ship production and possible economic multiplier–driven increases in economic activity from such indigenous production.

Newport News Shipbuilding Case Study

Newport News Shipbuilding (NNS) is the largest private sector single-site employer in the Commonwealth of Virginia and is a major economic engine of the Hampton Roads region of the United States. RAND's examination utilized extensive interviews with subject-matter experts, open literature, and publicly available data.

NNS is an "employer of choice" in its region. NNS pays its employees well and has only limited annual attrition. Shipbuilder jobs tend to be considerably more desirable than most workers' next-best alternative, especially in light of most workers' reluctance to relocate.

NNS appears to have generated relatively few local spin-offs. Experts are concerned that the Hampton Roads region, in general, lacks a heritage of entrepreneurial behavior.

The area immediately proximate to NNS is not economically vibrant. Experts told us that NNS workers rush to their cars at the end of their daily shift (15:30) and leave the vicinity as expeditiously as possible.

Austal USA Shipbuilding Case Study

Whereas NNS is a long-established shipbuilder, Austal operations in Mobile, Alabama, only developed in earnest in the last ten years. Austal USA's scale of operation increased by nearly a factor of five between 2009 and 2014 (though it remains considerably smaller than NNS both in terms of revenue and employment level).

Most Austal employees live in Mobile County or nearby Baldwin County, Alabama, but a sizeable fraction commute from neighboring areas of the states of Mississippi and Florida. Given that shipbuilder jobs are both unique and relatively well paying, we have consistently found a willingness on the part of shipbuilder employees to undertake significant driving commutes.

In order to obtain required training for its growing workforce, Austal USA has relied on the State of Alabama–funded Maritime Training Center. Between individuals trained at the Maritime Training Center not ultimately hired by Austal and considerable attrition at Austal USA, Austal has sizably altered the workforce-skill profile in the greater Mobile area beyond its current employees. Austal has not (at least yet) caused the development of a network of proximate local suppliers.

The Gripen Case Study

Sweden's JAS-39 Gripen fighter program has been lauded for successfully delivering an advanced fighter aircraft while also producing a significant economic multiplier to the local and national economy. It has been extensively cited in discussions of Australia's shipbuilding industry (e.g., Roos, 2014, Economic Development Board South Australia, 2014). The RAND research team therefore conducted a literature review and interviewed subject-matter experts to examine the Gripen program's wider benefit to Sweden.

Beginning in the early 1980s, the Gripen aircraft was produced by Saab in Linkoping in central Sweden, about 170 kilometers to the southwest of Stockholm. The program originally had a target for the creation of 800 jobs in a region with high unemployment. By 1987, the program had generated an estimated 1,200 new jobs. Today, the program is thought to sustain roughly 3,000 jobs in Sweden, with hopes to market a "next generation" upgrade of the Gripen through to 2040. Anchored around Saab, the local technical university, and a number of science parks, the wider Linkoping aerospace cluster currently employs approximately 18,000 workers, approximately one-third of the local workforce. A number of academics, most notably Eliasson (2010, 2011), have argued that the program has generated significant knowledge spillovers and a variety of spin-off firms, several working in areas quite distant from aviation. The program is also credited with helping to sustain such established firms as Volvo and Ericsson. The Gripen program appears to have had a larger (more favorable) economic multiplier, estimated by Eliasson (2010) to be around 3.6, than the 1.7–1.9 range more typically found for major defense projects.

Discussion

In terms of what individuals employed in Australian shipbuilding would be doing otherwise, key issues are the state of the Australian (and the shipbuilder's regional) economy and the degree of difficulty that shipbuilding workers would have finding commensurate alternative employment.

Our examination of shipbuilders in the United States finds slack economies in Mobile County and Hampton Roads and considerable rigidity in workers' abilities to find commensurate employment. Workers employed in shipbuilding appear to be quite geographically immobile (though willing to incur sizeable driving commutes). Both NNS and Austal USA are able to attract many job applicants, suggesting that these workers do not have alternative employment options as desirable as working for the shipbuilders. Several experts noted a tendency for laid-off shipbuilding workers to have prolonged periods of unemployment or underemployment, awaiting recall to the shipbuilder. The shipbuilders have not displaced high-value activities for many of their workers.

Regarding the extent to which shipbuilding in Australia would generate favorable spin-offs and spillovers, the U.S. examples are not remotely as optimistic as the Gripen example. For example, NNS appears to have generated relatively few spillovers. Indeed, the entire Hampton Roads region has been critiqued for a dearth of entrepreneurial activity. Likewise, no cluster of suppliers has yet emerged around Austal USA. The Gripen analogy appears to be overly optimistic as to the magnitude and nature of spin-offs and spillovers that might be expected from naval shipbuilding in Australia.

The indigenous production of ships in Australia cannot be expected to have both low opportunity costs and displacements and high levels of favorable spillovers. Indeed, we believe that these two objectives trade off against one another.

Acknowledgments

The RAND research team benefited from discussions with Carl-Henrik Arvidsson, Scott Benson, Audra Caler-Bell, Bo Carlsson, Russell Chandler, Semoon Chang, Charles Colgan, Bill Docalovich, Irwin Edenzon, Gunnar Eliasson, Hugh Green, Dan Gulling, Sten Gunnar Johansson, Donald Keeler, Magnus Klofsten, Jeffrey Kobrock, James Koch, David Lambert, Brian Leathers, Hugh Lessig, Craig Perciavalle, William Pfister, Yvonne Rosmark, Stephen Smith, Kjell Sullivan, Michelle Tomaszewski, Gary Wagner, and Troy Wayman. We also appreciate the assistance we received from Phillip Dolan, John Dykema, Chris Mather, Matthew Mulherin, and Benjamin Zycher.

Kate Louis and Glenn Alcock from Australia's Department of Defence provided a number of useful comments on an earlier version of this report. We also appreciate the assistance we received from Marc Ablong and Cameron Gill.

Cynthia Cook and Paul DeLuca provided constructive program-level oversight. Charles Colgan, director of research, Center for the Blue Economy, Middlebury Institute of International Studies at Monterey, and James Hosek of RAND provided reviews of an earlier version of this report. Judith Mele assisted us with shipbuilder contract value data. Lisa Miyashiro created the maps presented in this report. Matthew Byrd orchestrated the document publication process; Rebecca Fowler edited this report. We also thank our RAND colleagues Daniel Abramson, Philip Anton, Mark Arena, John Birkler, Olivia Cao, Robert Case, Christopher Dirks, James Kallimani, Krishna Kumar, Gordon Lee, Roger Lough, Kimbria McCarty, Nancy Moore, Eric Peltz, Ellen Pint, Hans Pung, John Raffensperger, John Schank, and Cathy Zebron for their assistance and insights.

CHAPTER ONE

Introduction

This report on the economic consequences of shipbuilding is part of a larger RAND project for the Australian government entitled Analysis of Australian Shipbuilding Industry and Capabilities. The larger project is to inform Australian policymakers of the economics and feasibility of various strategies for the Australian shipbuilding industrial bases that produce or repair naval surface vessels.

Other products from this project will provide much greater detail on prospective strategies that the Australian government might follow. For purposes of this report, however, we distill options to two corner solutions, and we offer a third approach that hybridizes the two corner solutions.

One approach, emulating the approach of the U.S. government, would be for the Australian government to pay Australia-based shipbuilders to build all Australian naval vessels. These Australia-based shipbuilders would also be responsible for acquiring and integrating the weapon systems used on the ships.[1]

The opposite approach would be for the Australian government to acquire ships for the Australian Navy from foreign providers. The ships could be purchased through a government-to-government transaction (akin to the U.S. Department of Defense's Foreign Military Sales program), or the Australian government could have a direct contractual arrangement with a foreign shipbuilder. Irrespective of the exact legal

---

[1] The firms that produce equipment used on the ships may or may not be Australian. Even U.S. Navy ships have some internationally produced equipment. The prime contractor ultimately responsible for the ship's performance is indigenous under this approach.

arrangement, the Australian government would send funding to a foreign government or shipbuilder, and that foreign entity would provide a ship to the Australian Navy. Under this type of arrangement, Australian firms would not participate except if serving (under mandate in the arrangement or at the foreign shipbuilder's discretion) as subcontractors to the foreign shipbuilder.

These two corner solutions (build in Australia or purchase ships abroad) could be hybridized in a number of ways. The Australian government, for example, could decide to purchase certain types or classes of ships abroad while having indigenous production of other types or classes of ships. Another hybridization would be for the Australian government to purchase partially completed ships abroad and complete them in Australia. For instance, Australia might purchase a ship that can float and operate under its own power, but the installation of its weapon systems could be done in Australia.

A panoply of technical, industrial base, diplomatic, and military issues arises in considering the two corner solutions, as well as the hybridization option. This report, however, has the narrower task of assessing the economic consequences of these options. We categorize these prospective economic consequences under two rubrics:

1. **Opportunity cost/displacement.** The basic question here is what individuals who would be employed in shipbuilding in Australia would be doing if that work were not in Australia. The opportunity cost and displacement could be very low (e.g., the worker would otherwise be unemployed) or quite considerable (e.g., the worker would be employed in a different type of manufacturing in Australia, but that alternative manufacturing would be priced out of existence in Australia if it had to compete with shipbuilding for skilled labor). There is likewise an opportunity cost associated with spending by the Australian government. If purchasing a ship abroad saved money, the Australian government could use the funding elsewhere or reduce taxation in the country.

2. **Spin-offs/spillovers.** We use the term *spin-off* to refer to a new firm that spins off an established firm—for example, a group of

employees decides to leave and set up a new business. A *spillover* is a more general expression for a positive, perhaps unanticipated, consequence of a project, such as the firm being able to enter a different industry or trained workers moving to other employers. Shipbuilding in Australia may have spin-offs and spillovers into the broader Australian economy.

Table 1.1 presents the broad approaches and questions related to opportunity cost and displacement and spin-offs and spillovers that arise in the context of each option. One might view this report as an attempt to address some of the queries presented in Table 1.1.

To address the questions in Table 1.1, this report uses analogies from the United States and Sweden to draw insights about the economic consequences of shipbuilding in Australia. To inform and prepare for our case studies, Chapter Two reviews the extensive literature on economic multipliers. This literature suggests considerable uncertainty as to the change in economic activity that will result from a shipbuilding project. Impacts could be quite favorable (i.e., a large multiplier) if shipbuilding gives rise to a cluster effect of firms benefiting

Table 1.1
Australian Shipbuilding Alternatives and Questions About Economic-Consequence Rubrics

Approach	Opportunity Cost/Displacement	Spin-Offs/Spillovers
Indigenous production	What would individuals employed in Australian shipbuilding be doing otherwise?	To what extent would shipbuilding in Australia generate favorable spin-offs and spillovers?
Purchase abroad	Assuming that foreign-built ships cost less, what would the Australian government and/or taxpayers do with cost savings?	Could the Australian government or Australian taxpayers invest cost savings in a realm that generates more-favorable spin-offs and spillovers?
Hybridization (buy some ships or some parts of ships abroad)	Which skill sets would be required in Australia under the hybridization, and what would be the opportunity cost and displacement associated with these workers?	Are there ways to structure the hybridization to maximize the extent of favorable spin-offs and spillovers?

from one another in a region. Net impacts, on the other hand, could be quite minimal (i.e., a multiplier near 0) if shipbuilding simply displaces other economic activities. Chapter Two also further discusses the economic trade-offs associated with indigenous production versus purchasing ships abroad.

Chapters Three through Five present three different case studies:

- Chapter Three: Newport News Shipbuilding (NNS) in Newport News, Virginia
- Chapter Four: Austal USA shipbuilding in Mobile, Alabama
- Chapter Five: The Gripen program undertaken by Saab Aeronautics in Linkoping, Sweden.

We chose these three case studies for different reasons.[2] NNS is the largest shipbuilding company in the United States. It has received sizable funding for many years, and therefore represents a mature case. Whatever impacts NNS has had on its region should be observable. By contrast, Austal USA is new to the United States, with operations in Mobile commencing in earnest within the last ten years. Hence, Austal provided new-shipbuilder insights that NNS could not provide. The Gripen program, meanwhile, came to our attention because it has been discussed considerably in the context of shipbuilding in Australia (e.g., Roos, 2014, Economic Development Board South Australia, 2014).

A case study methodology of the sort we use in Chapters Three to Five has advantages and disadvantages. For each of our three cases, RAND researchers traveled to the locations to see what has happened. We were also able to conduct in-depth, on-site interviews with subject-matter experts. Some of these experts were employed by the firms involved, but others were experts on the local economies not affiliated

---

[2]   We additionally conducted interviews and a literature review related to the Bath Iron Works in Bath, Maine; Ingalls Shipbuilding in Pascagoula, Mississippi; and Marinette Marine Corporation in Marinette, Wisconsin. Our examinations of these shipbuilders did not extend to the same level of depth as our case studies. However, we drew inferences and insights from these additional examples to complement those we drew from our formal case studies.

with the firms (e.g., local economics professors, economic development authority personnel). These interviews provided rich insights into local economic conditions and the impacts of these major firms.

A concern with a case study methodology is that one is gathering anecdotes, not data. There is validity to this concern. Each case has its own idiosyncrasies that obviously affect its outcomes. However, an examination such as this will never be able to draw on a large data set—there simply are not that many places in the United States or other industrialized democracies where military ships are built.

There was an additional data challenge that we faced: This project's Australian sponsor is a foreign government, both from the perspectives of U.S. shipbuilders and Saab in Sweden. We were not in a position to make detailed data-related demands on our hosts. Naturally, firms had incentive to present themselves favorably, but we also had access to government data sources and disinterested experts and literature to corroborate or refute what we were told.

Analyses of impacts of shipbuilding in Australia, such as the analysis by Deloitte Access Economics (2014) of Australia's experience with the *Collins* submarine program, can draw on more-granular and more-detailed information about where, for example, the shipbuilder spent money. Of course, as there are a limited number of examples of military shipbuilding in the United States, there are yet-fewer examples in Australia.

Our case studies are regionally, not nationally, focused. Analysis of the national effects of shipbuilding could yield different findings. One possibility is that a shipbuilding project could draw workers away from other regions, to those regions' detriments, though we find little evidence for such an effect, especially not for blue-collar workers. Another possibility is that the shipbuilding project could invigorate other regions where suppliers are based. Our thinking, however, is that maximum effects are likely to be observed at the shipbuilder's home location, so our case studies present upper-bound estimates of the possible economic consequences of investing in shipbuilding.

Chapter Six presents a discussion that synthesizes the findings of the preceding chapters.

CHAPTER TWO

Economic Multipliers and Their Implications

This chapter provides an overview of the extensive literature on economic multipliers.[1] While the literature is sizable, its implications for the economic consequences of shipbuilding in Australia are uncertain. Multiplier estimates vary widely, both in magnitude and sign. The economic multiplier that will apply to a given project varies with the economic conditions and specific context in which the shipbuilding spending would occur.

The magnitude of the economic multiplier from shipbuilding prominently figures in the economic trade-off that the Australian government faces between indigenous production and purchasing ships abroad.

The Logic of Economic Multipliers

An economic multiplier is a basic concept in macroeconomics. Suppose that the government spends $100 buying a good or service from a company or individual. The party that receives the $100 might then be expected to spend the $100 (or at least a portion of it) on some other goods and services. The original $100 creates a cascade (i.e., multiples)

---

[1] Interestingly, and advantageously from our perspective, the broader macroeconomic literature often focuses on defense spending examples. The big advantage of defense spending, from a macroeconomic researcher's perspective, is that its funding patterns are plausibly exogenous. Military exigency, not the Great Depression, led to the massive government spending associated with the Second World War, for instance. See Barro and Redlick (2011).

of spending through the economy. That $100 spent at a shipbuilder results in additional spending by shipbuilder workers at local restaurants, which then hire additional workers who rent additional housing, and so forth.

The cascade of resultant spending is not going to be infinite, because recipients at each stage may save some of the money, will have to pay taxes on it, and may spend some of the money outside the measured region (outside the country, e.g., imported goods, applies to a national-level multiplier; outside the region being studied applies to a state- or local-level analysis). If a leakage (saving, taxation, imports) rate can be estimated, the economic multiplier represents the sum of a convergent geometric series. For example, with a leakage rate of 50 percent, the original $100 in spending would result in a total regionwide spending increase of $200 ($100 plus $50 plus $25 plus $12.50 plus $6.25, and so on), or an economic multiplier of 2.0. Note that multiplier values in this formulation are inclusive of the originating level of spending.

Many studies have estimated economic multipliers associated with defense spending. Arena, Stough, and Trice (1996) estimated a personal income multiplier of 1.7 for spending at NNS. Ironfield (2000, 2002) finds national output multipliers of 1.95 for her analyses of both the ANZAC ship and the Minehunter Coastal Project. Hosek, Litovitz, and Resnick (2011) find multipliers of 1.83 (payments to defense-employed personnel) and 1.95 (payments for defense procurement) for spending in Hawaii. Acil Allen Consulting (2013) found an Australian ship production gross domestic product multiplier of 1.78. The U.S. Maritime Administration (2013) asserted a multiplier of 3.66, perhaps because the analysis was for the United States—nationwide— rather than the regionally focused analyses of Arena, Stough, and Trice (1996) and Hosek, Litovitz, and Resnick (2011).

Economic multipliers are estimated using input-output modeling. Different categories of expenditures are established, and it is estimated where a dollar, if spent in a category, is subsequently spent across other categories. The calculations are, in a sense, a bookkeeping-type exercise. A dollar sent to a recipient (e.g., a shipbuilder) is divided between categories of usage (e.g., a share to pay workers, a share to purchase materials from subcontractors, a share paid in dividends to shareholders, a share paid in taxes), then each category, in turn, has its own allocation of consequent spending.

Multiplier estimates rely on knowing the categories and estimating the magnitudes of producer spending. If the technology to be used in shipbuilding is unknown or different from analogies, multiplier estimates derived from analogies may be inaccurate.

Hosek, Litovitz, and Resnick (2011) used a U.S. Bureau of Economic Analysis input-output methodology called Regional Input-Output Modeling System (RIMS). Arena, Stough, and Trice (1996) used a similar approach called Impact Analysis for Planning (IMPLAN). The U.S. Maritime Administration (2013) used the Minnesota IMPLAN Group (MIG). Speaking specifically of IMPLAN (but also applying to RIMS and MIG), Morgan (2010, p. 5), notes:

> IMPLAN is a static model and does not capture the dynamics of how a regional economy might change over time. It assumes, as most standard input-output models do, that wage levels, prices, property values, input costs, labor supply, productivity, and other key variables will remain constant. As such, IMPLAN is not readily suitable for forecasting the effects of public policy changes.

Likewise, Hosek, Litovitz, and Resnick (2011, p. 16) warn readers:

> Although the model offers a cohesive framework for viewing these flows and their relationship to final demand, this is not the same as identifying the underlying structural relationships or showing the causal effect of a given change in demand or production. The model is not designed to estimate the effect of changes in defense spending on the economy.

Interpretive Limitations of Economic Multipliers

Unfortunately, building on comments from Morgan (2010) and Hosek, Litovitz, and Resnick (2011), estimated economic multipliers do not necessarily lend themselves to a plausible and desired usage of the estimate. Just because one has estimated a multiplier of, say, 2.0, it does not logically follow that a $100 million government expenditure on shipbuilding in Australia will result in an increase in economic activity of $200 million.[2]

An example from the Second World War illustrates the problem. The vast defense-related buildup that the United States' manufacturing base undertook came at the expense of the production of consumer products. Herman (2012, p. 153) notes:

> First came the auto industry, with a drastic cut by more than half. Then in October [1941] nonessential construction was ordered halted, to divert materials to defense plant construction. On October 21 manufacturers had to stop using copper in almost all civilian products, followed by sharp cuts in refrigerators, vacuum cleaners, metal office furniture, and similar durable goods.

The total amount of economic activity generated by U.S. manufacturers did not increase by the magnitude of their government defense contracts. Rather, the net change was the value of those contracts less the value of the consumer products that would have been manufactured absent the government contracts.

This displacement or crowding-out effect has led other analysts to estimate much lower defense spending multipliers—e.g., 0.6 (Hall, 1986), 0.5 (Hall, 2009), and 0.4–0.5 within the first year and 0.6–0.7 over two years (Barro and Redlick, 2011). Cohen, Coval, and Malloy (2011, p. 2) also find "strong and widespread evidence of corporate retrenchment in response to government spending shocks"—that is, government spending crowding out private sector spending. Deloitte Access Economics (2014) used a computable general equilibrium approach that, because of a full-employment assumption, results in an economic multiplier estimate near zero for the *Collins* program.

---

[2] A multiplier refers to how a change in government spending (e.g., a new shipbuilding contract) changes gross domestic product or some comparable measure of total economic activity. If the multiplier is 1.0, economic activity increases by the dollar value of the contract, but no more. A multiplier can be zero if the shipbuilding contract completely displaces private sector activity, netting no change in total economic activity. If one considers tax effects and reduced economic activity from increased taxation, a government program could have a negative multiplier.

On the other hand, Fisher and Peters (2010) and Nakamura and Steinsson (2013) find multipliers of about 1.5, closer in magnitude to the Arena, Stough, and Trice (1996), Ironfield (2000), Ironfield (2002), Hosek, Litovitz, and Resnick (2011), and Acil Allen Consulting (2013) estimates. Nakamura and Steinsson (2013, p. 2) note that "there is no 'single' government spending multiplier." Rather, it varies with the economic conditions and the specific context in which the government spending occurs.

A key question, therefore, is what activities are displaced by the government's shipbuilding contract. The higher the economic value of the displaced activity, the lower the net economic benefit of the shipbuilding contract would be. Or, on the other hand, if the project results in a commensurate reduction in unemployment, a greater multiplier would be found.

It would be logical to hypothesize that an economic multiplier would be higher when there are a larger number of unemployed workers and therefore less displacement and a lower opportunity cost associated with increased defense spending. However, Hall (2009, p. 11) finds: "World War II does not yield a higher estimate of the multiplier than does the Korean War, despite the fact that the buildup starting in 1940 was from a much more slack economy than was the one starting in 1950." Owyang, Ramey, and Zubairy (2013) present the nonintuitive finding that multipliers are not higher during times of slack in the United States, but, for Canada, they find evidence for multipliers that are substantially higher during periods of slack in the economy.

An additional issue is whether and to what extent an analysis should consider the tax increases required to finance an increase in defense spending. If the decision has already been made to invest in a product and the only uncertainty is whether to purchase it domestically or internationally, increased taxes to pay for the product are not relevant unless there is a price premium for the domestically produced product. Also, if the multiplier analysis focuses on a small city or region in a much larger country (e.g., Hawaii in the United States), the local effect of increased taxes is likely to be negligible. But a national-level multiplier analysis that also considers increased taxation to fund increased defense spending could result in a negative multiplier, because of the excess burden or distortion associated with generating tax revenue. Or, in the other direction, reduced defense spending could have favorable economic consequences when one considers a concomitant reduced level of taxation (see Barro and Redlick, 2011, and Zycher, 2012).

It is also possible that estimated economic multipliers understate long-run economic benefits. Suppose, for instance, that the shipbuilding contract leads to skill development in a region that consequently attracts other industry. Then the economic implications would be larger than those estimated by an economic multiplier. Manski (2013), for instance, discusses the possibility of public spending enhancing private productivity. The "leakage arithmetic" of economic multiplier calculations does not consider the possibility of favorable economic spillovers or synergies. Such an effect would argue for use of a larger value as the applicable economic multiplier.

The near-zero multiplier from Deloitte Access Economics (2014) presents the argument that shipbuilding has no impact. The contrary argument is that shipbuilding has large-scale beneficial effects (i.e., a large multiplier), as espoused by Roos (2014) and Economic Development Board South Australia (2014). The latter argument is built on skill development and technology externalities. We next discuss three theories of technological externalities presented in the economics literature.

Technological Externalities

A technological externality exists when innovations and improvements in one firm increase the productivity of other firms without full compensation (Glaeser et al., 1992). We present three different theories of externalities:

- The Marshall-Arrow-Romer (MAR) theory says that the concentration of firms in the same industry in a city helps knowledge spillovers between firms, facilitating innovation and growth (Carlino, 2001). Examples include the semiconductor and software industries in the San Francisco Bay Area's Silicon Valley and the automotive industry in Detroit, Michigan (see Klepper,

2010). The MAR theory notes an advantage to local monopoly because a "local monopoly restricts the flow of ideas to others and allows the externalities to be internalized by the innovator" (Glaeser et al., 1992, p. 1127).

- Porter's theory also says that the concentration of an industry helps knowledge spillovers (1990). Examples include the Italian ceramics and German printing industries.
- Jacobs's theory says that knowledge transfers from outside an industry have greater impact than knowledge transfers within an industry (1969). Jacobs also argues that the rate of innovation with competitive markets was greater than the rate of innovation with monopolies.

Feldman and Audretsch (1999) find evidence that diversity of economic activities in a region (i.e., a region that hosts a variety of different industries rather than specializing in a specific industrial niche) promotes technological change and subsequent economic growth, and they find little support for specialization generating innovative activity. Glaeser et al. (1992) find that industrially diversified areas grew faster than specialized areas. On the other hand, Carlino (2001) finds little evidence that industrial diversity was an important factor in determining the rate of patenting activity in the 1990s.

One scenario that created clustering and even larger economic multipliers was spin-offs. According to Klepper (2010, p. 16): "With spinoffs not venturing far from their geographic origins, this led to a buildup of superior firms in Detroit and Silicon Valley." However, conditions that are conducive to spin-offs are unknown, and while Silicon Valley continues to experience growth, the Detroit area has declined considerably in recent years.

Economic Trade-Off

Australia faces an economic trade-off associated with indigenous production versus purchasing ships abroad; in this situation, the economic multiplier associated with shipbuilding is a key parameter.

Suppose Australia faces a choice between buying a ship from a foreign shipbuilder for $1 and paying a domestic shipbuilder $(1 + p)$ for the ship. In this case, p is the percentage price premium (or, if negative, discount) associated with indigenous production. For illustrative purposes here, we assume that the foreign-produced and indigenous options are equivalent from a capability perspective.

In order to purchase the foreign ship, the Australian government must generate $1 in tax revenue. However, as noted by Barro and Redlick (2011) and Zycher (2012), there are distortions and costs imposed from taxation, so the societal loss to generate $1 in tax revenue would be some larger amount $(1 + L)$. Taxation can decrease private production, decrease consumer and producer surplus, and induce behavior to evade the tax. Zycher (2012) discusses estimates of L of 0.35–0.5.

If the indigenous ship is purchased, the Australian government must generate $(1 + p)$ in tax revenue. With distortion from taxation, the societal cost of indigenous production would be $(1 + L)(1 + p)$.

However, indigenous production may also be associated with an increase in economic activity from the shipbuilding activity, $m_s(1 + p)$, where m_s is the economic multiplier associated with shipbuilding.

Table 2.1 summarizes the trade-off. The challenging scenario in Table 2.1 is when both $m_s > 0$ and $p > 0$. Then the decisionmaker faces a trade-off between increased tax burden from indigenous production and increased economic activity from that production.

As the indigenous price premium, p, increases, the relative cost of indigenous production increases. But so, too, does the local level of economic activity associated with indigenous production. As L, the level of distortions and costs associated with taxation, increases, the relative cost of indigenous production increases (with no increase in

Table 2.1
Trade-Offs Associated with Indigenous Production Versus Foreign Purchase

Approach	Tax Cost to Society ($)	Change in Economic Activity
Indigenous production	$(1 + L)(1 + p)$	$m_s (1 + p)$
Purchase abroad	$(1 + L)$	—

economic activity), assuming $p > 0$. As m_s, the economic multiplier associated with shipbuilding, increases, the change in economic activity from indigenous production increases.

The columns in Table 2.1 are not directly comparable or additive. The tax cost column denotes an actual cost to society, representing the payment of taxes plus the distortions inherent in taxation. The column about the change in economic activity only partially denotes a benefit to society. A shipbuilder worker, for example, must expend often physically demanding effort and forfeit leisure as part of his or her job. Hence, the wage he or she receives is not a pure net benefit. Rather, it is a gross payment—only a portion of which is a net benefit to the worker.

Another possibility would for the Australian government to purchase the ship abroad but not change its taxation level to offset the cost savings. In this scenario, the Australian government would have an additional p available to spend in other areas it felt worthwhile. Under this *displaced spending* argument, the opportunity cost of indigenous production would not be in the form of increased taxes but rather in the form of reduced, socially beneficial government spending.

Table 2.1's tabulation makes several key assumptions:

- Capabilities are the same across the two approaches.
- The ship purchased abroad has no Australian content (i.e., from Australia-based suppliers). If that ship had Australian content, it would have a multiplier value in the "Change in Economic Activity" column.
- The multiplier m_s is scaled in proportion to the indigenous ship's level of Australian content. To the extent that the indigenous ship uses non-Australian inputs, m_s would be lower. (Of course, if the two approaches had the same level of Australian content, they would presumably have the same multiplier value.)
- Australia's choice of indigenous versus foreign production does not affect other Australian firms (e.g., the ability of Australian firms to export to foreign markets).

Key Issues for Our Case Studies

This chapter has highlighted some key issues in estimating the shipbuilding economic multiplier that we will explore in our case studies. These issues include

- What were shipbuilder workers doing before they were hired by the shipbuilder? What might they plausibly be doing if not employed by the shipbuilder? To the extent that shipbuilding reduces unemployment or underemployment, a larger m_s is implied. If shipbuilding displaces comparable private sector work, a low or zero m_s is implied.
- Has the shipbuilder lost skilled employees to other firms in the region? High attrition to other employers would have mixed implications with respect to m_s. On one hand, it might suggest that a shipbuilding job is not much better than or different from what the worker might otherwise be doing, consistent with a low value of m_s. On the other hand, shipbuilder attrition could be consistent with favorable economic spillovers (e.g., the shipbuilder trains workers who then enhance private sector economic development in the region).
- Has the shipbuilder generated spin-off firms undertaking nonnautical or nondefense business? If such a phenomenon were widespread, it would suggest that the economic benefit of shipbuilding is understated by economic multiplier arithmetic.

We begin our case studies in the next chapter by analyzing NNS.

CHAPTER THREE

Newport News Shipbuilding Case Study

Background on Newport News Shipbuilding

NNS, a division of Huntington Ingalls Industries, is the largest ship-building company in the United States.[1] Founded as the Chesapeake Dry Dock and Construction Company in 1886, NNS is the largest industrial employer in the Commonwealth of Virginia.[2] NNS has built more than 800 naval and commercial ships; however, since the end of the 1990s, NNS has exclusively built ships for the U.S. Navy. NNS is the United States' sole designer, builder, and refueler of nuclear-powered aircraft carriers, as well as one of two shipbuilders (the other being Electric Boat in Groton, Connecticut) with the capability to design and build nuclear-powered submarines. NNS built the world's first nuclear-powered aircraft carrier, the USS *Enterprise*, and all ten of the *Nimitz*-class nuclear aircraft carriers and is currently building the *Ford*-class nuclear aircraft carriers and *Virginia*-class nuclear-powered submarines.

In conjunction with its unique role in the production of nuclear-powered ships, NNS provides technical support for these vessels. It is the only shipyard in the United States that can undertake the midlife refueling complex overhaul (RCOH) of a nuclear aircraft carrier—a task that it has begun for the *Nimitz* class. An RCOH is a signifi-cant endeavor; the refueling of the USS *Nimitz*, completed in 2001,

---

[1]   This paragraph and the next draw on Huntington Ingalls Industries (2014) and NNS (n.d.).

[2]   Virginia is officially a commonwealth. This term, in U.S. usage, is a synonym for *state*.

took five years of planning and three years of execution. Schank et al. (2002, p. xiii) described carrier RCOH as one of "the most challenging engineering and industrial tasks undertaken anywhere by any organization."

The U.S. government's web site USASpending.gov (n.d.) provides some insight about the volume of U.S. government funding flowing into NNS. Figure 3.1 shows annual obligations between fiscal years 2000 and 2013.[3] Figure 3.1 is, however, misleadingly jagged: The shipbuilder signed several large contracts in fiscal year 2001, but the actual government funding did not flow to NNS until work was undertaken a year or more later. On average, however, Figure 3.1 shows that NNS has received obligations of roughly USD 3 billion per year from the U.S. government. But not all of these obligations were spent in the region; NNS has a network of suppliers throughout the United States.

Figure 3.1
U.S. Government Contractual Obligations to Newport News Shipbuilding, Fiscal Years 2000–2013

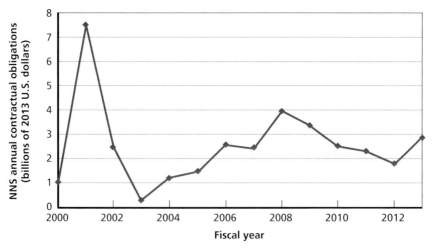

SOURCE: USASpending.gov, n.d.

RAND RR1036-3.1

---

[3]   A U.S. government fiscal year runs from October 1 to September 30. Fiscal year 2000, for instance, ran between October 1, 1999, and September 30, 2000.

NNS is located on a 2.2-million-square-meter site in Newport News, a city with about 182,000 residents. The city runs on a northwest-to-southeast angle along the James River waterfront; NNS is located at the southern end of the city. The city of Newport News is part of the Virginia Beach–Norfolk–Newport News Metropolitan Statistical Area, a region with a population of about 1.7 million people. The metropolitan area is built around the body of water known as Hampton Roads, one of the world's largest natural harbors. See Figure 3.2. The famous American Civil War naval battle between the USS *Monitor* and the CSS *Virginia* (formerly USS *Merrimack*), two iron-clad ships, took place in Hampton Roads on March 8–9, 1862, not far from where NNS is now located. The Port of Hampton Roads is the deepest and third-largest port on the East Coast of the United States (Norfolk Department of Development, n.d.).

The Economic Consequences of Newport News Shipbuilding

As mentioned in Chapter Two, Arena, Stough, and Trice (1996) estimated a personal income multiplier of 1.7 for spending at NNS. While that analysis is nearly 20 years old, no one we spoke to, either affiliated with NNS or not, felt that underlying conditions in the Hampton Roads region have changed markedly. NNS was and remains the largest private sector single-site employer in Virginia and a major economic engine of the Hampton Roads region.

While the Hampton Roads region is home to other defense-related entities—including Langley Air Force Base, Norfolk Naval Air Station, and the Norfolk Naval Shipyard (NNSY)—NNS seems to be largely disconnected from these other entities, with the possible exception of NNSY. These other defense-related entities' existence in the region does not imply that it has the type of salubrious clustering that we discuss in the Gripen example in Chapter Five. Rather, these colocations may, in large part, reflect the desirable geographic location.

The U.S. Department of Labor, Bureau of Labor Statistics (n.d.) reports a 6.2-percent unemployment rate for the city of Newport

Figure 3.2
The Hampton Roads Region

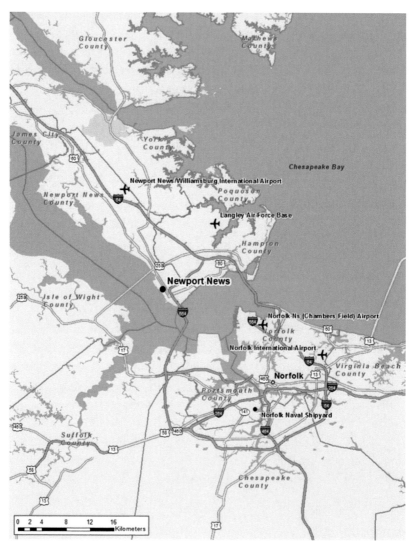

SOURCE: Map created using ArcGIS® software by Esri, with TeleAtlas as the data source. Copyright Esri.

News as of September 2014. The Virginia Beach–Norfolk–Newport News Metropolitan Statistical Area had a 5.6-percent unemployment rate that month. The Commonwealth of Virginia's unemployment rate was 5.2 percent.

As of our analysis, NNS employed approximately 24,000 workers. NNS draws its workers from a broad swath of the region. An NNS expert estimated that on the order of 1,000 employees live in North Carolina, driving more than 70 kilometers to NNS. Consistent with this observation, Table 3.1 presents U.S. Census 2011 data on where individuals whose primary job was in the private sector in Newport News lived. Of course, NNS is, by a wide margin, the largest private sector employer in Newport News, representing about 30 percent of the city's private sector employment.[4]

As is true of other shipbuilders we have examined, NNS is an "employer of choice" in its region. NNS pays its employees well and has only limited attrition (estimated at 3–5 percent annually). NNS job openings elicit a large number of applicants. The only other employer in the region that is comparable in terms of size and skill requirements is the U.S. Navy's NNSY, located in Portsmouth, Virginia. The NNSY employs about 9,500 workers (McCabe, 2014). The NNSY, one of the United States' four naval shipyards, is tasked with the maintenance, repair, modernization, inactivation, and disposal of U.S. Navy ships and systems. Currently, the NNSY primarily services nuclear aircraft carriers and nuclear submarines (see Naval Sea Systems Command, 2012). Key points of differentiation between the NNSY and NNS are that the NNSY does not build ships and does not undertake RCOHs. Also, NNSY workers are employees of the U.S. government, whereas NNS is a private firm. The NNSY has not engaged in large-scale hiring in recent years, though it has plans for ramped up hiring in 2015 (McCabe, 2014). In recent years, the lackluster regional economy has allowed NNS considerable selectivity in its hiring, we were told.

---

[4]   In Chapter Four, we present data about where Austal USA workers live that we can then juxtapose with the type of U.S. Census data presented in Table 3.1. The Austal data suggest that U.S. Census data underestimate shipbuilder workers' commuting distances. We therefore view Table 3.1 as a lower bound on the residential dispersion of NNS's workforce.

Table 3.1
Residence of Private Sector Employees in Newport News, 2011

Residence	Number	Percentage
City of Newport News	23,459	29.4
City of Hampton	12,533	15.7
York County	6,006	7.5
City of Virginia Beach	3,935	4.9
City of Chesapeake	2,954	3.7
Gloucester County	2,915	3.6
Isle of Wight County	2,900	3.6
City of Suffolk	2,835	3.5
James City County	2,689	3.4
City of Norfolk	2,409	3.0
City of Portsmouth	2,120	2.7
City of Poquoson	1,332	1.7
Chesterfield County	812	1.0
Elsewhere in Virginia	10,481	13.1
North Carolina	1,677	2.1
Maryland	321	0.4
Other	489	0.6
Total	**79,867**	

SOURCE: U.S. Census Bureau, Center for Economic Studies, n.d., which uses Longitudinal Employer-Household Dynamics Origin-Destination Employment Statistics data (beginning of quarter employment, second quarter of 2011).

NOTE: The primary employment of these workers was in Newport News.

NNS has several accession pathways. NNS works with local community colleges to provide applicants with valuable skills (e.g., welding training). Additionally, for many years, NNS has had an apprentice school. NNS credits the apprentice school with providing many of the shipbuilder's leaders (see, for instance, Haun, 2014; Lessig, 2014; and Lessig, 2015). Mourshed, Farrell, and Barton (2013, p. 65) describe the apprentice school as a "huge cost saver for the company; by investing up front in acquiring talent, it saves down the line on expenses related to retraining and vacancies." A new building for the school was recently constructed near NNS; see Figure 3.3. Cooper (2013) notes that the Commonwealth of Virginia provided USD 25 million for the new building for the school, while the City of Newport News provided USD 17 million to buy the property and construct its garage and infrastructure. The developer Armada Hoffler invested USD 30 million for the residential and retail components, shown on the right of Figure 3.3.

For white-collar workers (NNS uses the vernacular *nonproduction*), NNS recruits information technology, computer science, and engineering graduates from around the United States. Experts told us that many of these white-collar workers—much more so than blue-

Figure 3.3
The Newport News Shipbuilding Apprentice School

SOURCE: Irina Danescu, 2014.
RAND *RR1036-3.3*

collar (*production*) workers—probably would not be living in the Hampton Roads region were it not for NNS; they would work and live elsewhere.

NNS also recruits extensively among former members of the U.S. military, for both production and nonproduction positions.

Echoing a point we heard at other shipbuilders, NNS has not hired many workers who previously worked at other U.S. Navy–funded shipbuilders. Rather, both in the context of NNS and other shipbuilders, we were told that shipbuilder workers are generally highly attached to their localities and reluctant to relocate (though willing to incur sizable daily driving commutes).

NNS has not had a trades-based layoff since 1999. However, NNS experts have a heuristic that they could rehire half of laid-off employees one year later, if they so desired. Subject-matter experts in other shipbuilder regions suggested an even greater ability to rehire laid-off employees. Shipbuilder jobs tend to be considerably more desirable than most workers' next-best alternative, especially in light of most workers' reluctance to relocate.

As mentioned, NNS has had limited attrition. As a result, while we heard anecdotes of NNS alumni (in several cases, retirees) setting up their own businesses, these anecdotes appear to be exceptions. We likewise heard an anecdote of a few workers who relocated from NNS to the Marinette Marine Corporation in Wisconsin, but the story was salient to the expert who relayed it exactly because of its unusual nature.

A broader concern raised by Koch (2014, slide 54) is that the Hampton Roads region lacks a heritage of entrepreneurial behavior (e.g., start-ups, spin-offs). In recent years, the region has ranked last among Virginia regions in per capita new business start-ups (see Bozick, 2014).

The area immediately proximate to NNS (depicted in Figure 3.4) is not economically vibrant (see Figure 3.5)—for example, there are a number of boarded-up buildings, a lack of busy businesses, and a nearly complete absence of pedestrians in the middle of the day.

Experts told us that NNS workers rush to their cars at the end of their daily shift (15:30) and leave the vicinity as expeditiously as pos-

Figure 3.4
The Immediate Vicinity of Newport News Shipbuilding

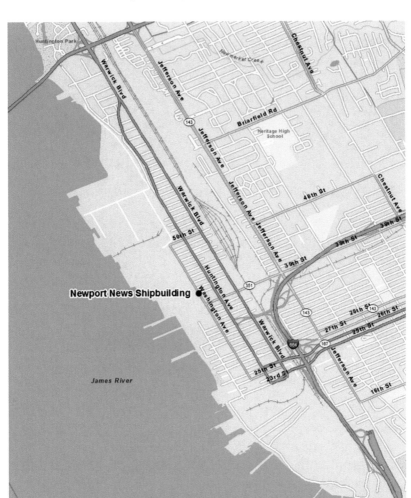

SOURCE: Map created using ArcGIS® software by Esri, with TeleAtlas as the data source. Copyright Esri.
RAND RR1036-3.4

Figure 3.5
A Lack of Economic Vibrancy near Newport News Shipbuilding

SOURCE: Edward G. Keating, 2014.
RAND *RR1036-3.5*

sible. Indeed, fee-based parking lots for NNS workers appear to be a major usage of property near the shipbuilder (see Figure 3.6).

The northern part of Newport News, farther away from NNS, is more economically vibrant. However, the Regional Studies Institute at Old Dominion University (2014) criticized a major hotel and conference center complex (see Figure 3.7) in the northern part of the city for being driven by public funding while not requiring any public accountability (e.g., any requirement for a minimum number of new jobs to be created).[5] It is unclear whether this facility would have been built absent a reported USD 26 million in governmental assistance.

---

[5]   The hotel's Newport News Marriott at City Center moniker is potentially misleading. The hotel is located in the northern part of the city, approximately 13 kilometers from the downtown area near NNS.

Figure 3.6
A Parking Lot near Newport News Shipbuilding

SOURCE: Irina Danescu, 2014.
RAND *RR1036-3.6*

Insights for Shipbuilding in Australia

Table 1.1 posed questions pertaining to shipbuilding in Australia. The NNS case study provides responses to some of those questions, as presented in Table 3.2.

Next, we discuss Austal USA shipbuilding in Mobile, Alabama.

Figure 3.7
Newport News Marriott at City Center

SOURCE: Irina Danescu, 2014.
RAND *RR1036-3.7*

Table 3.2
Insights for Australian Shipbuilding from the NNS Case Study

Opportunity Cost/Displacement	Insights from NNS Case Study
What would individuals employed by NNS be doing otherwise?	Subject-matter experts felt that NNS white-collar employees would likely live in other parts of the United States and probably would not work in the shipbuilding industry. NNS blue-collar workers might still live in the region, but some would have lower-skilled, lower-paid positions.
To what extent has NNS generated favorable spin-offs/spillovers?	The spin-offs and spillovers are very limited. While NNS workers have diffused income over a broad area, the entire Hampton Roads region has been criticized for a paucity of entrepreneurial activity. NNS has suppliers throughout the country without a pronounced local clustering. NNS has had low attrition.

CHAPTER FOUR

Austal USA Shipbuilding Case Study

Background on Austal USA Shipbuilding

Austal USA is the American branch of the Australia-based shipbuilder Austal. Its 670,000-square-meter site, opened in 1999, is located on Blakeley Island and adjoining Pinto Island in Mobile, Alabama.

When the facility opened, it produced commercial ships (including two ships for the now-defunct Hawaii Superferry).[1] These ships leveraged Australian Austal technology in building aluminium high-speed ferries using a catamaran design. Since 2009, Austal USA has shifted to solely designing and building U.S. Navy vessels; it has contracts for ten Joint High Speed Vessels (JHSVs), of which four have been delivered, and 12 Littoral Combat Ships (LCSs), of which two have been delivered. Austal USA's experiences with the two aluminum ferries clearly helped the firm win the JHSV and LCS contracts.

Whereas USASpending.gov provided insight on the volume of U.S. government funding flowing to NNS, this data source is less useful for Austal USA. Austal's first two LCSs were officially contracted to Bath Iron Works in Maine, with Austal USA as a subcontractor. Austal has only been the prime contractor from Austal's third LCS (the *Jackson*, LCS 6—Austal has built even-numbered LCSs) on, a requirement for contract obligation data to appear on USASpending.gov.

---

[1] Subsequent to the bankruptcy of Hawaii Superferry, the U.S. Navy ended up acquiring these ships. They are now the USNS *Guam* and the USNS *Puerto Rico*. See U.S. Department of Defense (2012).

A more informative (though still imperfect) way to understand Austal USA's growing level of business is to use Austal annual reports' presentations of revenue from operations in the United States. See Figure 4.1.

Figure 4.1 should be viewed as an approximation. Austal's annual reports present values in Australian dollars. We used annual January 1 U.S. dollar–Australian dollar exchange rates to translate the annual reports' July 1–June 30 revenue values into U.S. dollars.[2] Volatility in that exchange rate (which has been considerable) necessarily makes using a mid–fiscal year exchange rate an approximation.

There are two key points from Figure 4.1. First, Austal USA's scale of operation has escalated considerably, with revenue increasing

Figure 4.1
Estimated Austal USA Annual Revenue, 2009–2014

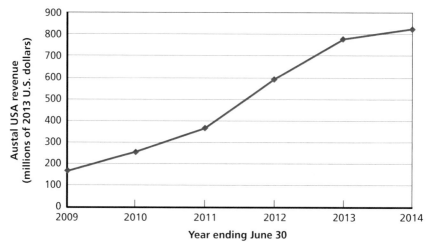

SOURCE: Austal Limited annual reports, 2009–2014.
RAND RR1036-4.1

2   For example, Austal Limited's annual report from 2010 states that Austal USA's revenue for that year was AUD 267 million. On January 1, 2010, USD 1.000 was worth about AUD 1.114 (per Oanda Corporation, n.d.). Hence, Austal USA's 2010 revenue was approximately USD 240 million. But then we used the gross domestic product price deflator from the U.S. Department of Commerce, Bureau of Economic Analysis (2014), to put Figure 4.1 into constant 2013 dollars, so the 2010 Austal USA revenue total is approximately USD 253 million (in 2013 U.S. dollars).

by nearly a factor of five between 2009 and 2014. Second, even at its 2014 revenue level, Austal USA's operations are considerably smaller than NNS's (about USD 800 million in Austal USA revenue versus roughly USD 3 billion in annual contractual obligations for NNS).[3] Likewise, Austal's current employment level of about 4,200, while markedly increased from about 800 in late 2009, is well below NNS's roughly 24,000.

Mobile, Alabama, is a city on the Gulf of Mexico with an estimated population of about 195,000 for 2013 (U.S. Census Bureau, 2014b). The broader Mobile metropolitan area has an estimated population of about 415,000 (U.S. Census Bureau, 2014a).

Austal USA's facility is located on Blakeley Island and adjoining Pinto Island, immediately to the east of downtown Mobile. The Mobile River separates the Austal facility from the downtown area; there are two tunnels and a bridge connecting the islands to the mainland. To the east of Austal is Mobile Bay, which is connected by two bridges to Baldwin County, Alabama. Figures 4.2 and 4.3 provide maps of the area at different scales.

Figure 4.4 shows a photo of the under-construction LCSs 6 and 8 in the Mobile River.

Figure 4.5 is a photograph of downtown Mobile, taken approximately where the photograph in Figure 4.4 was taken, but facing the opposite direction. Austal USA's shipyard is very proximate to downtown Mobile, but they are separated by the Mobile River.

In Chapter Three, we noted our concern with the economically challenged area adjacent to NNS. Austal USA's geography averts this problem to a considerable extent. There is no housing on Blakeley Island or Pinto Island; the islands are used for industrial purposes, as well as hosting the USS *Alabama* Memorial Park, which is east of the Austal facility on Blakeley Island. The Ingalls shipyard in nearby Pascagoula, Mississippi, has a similar arrangement, with the shipbuild-

---

[3]   There is clearly a timing difference between contractual obligations (see Figure 3.1) and revenue (see Figure 4.1). A contractual obligation should eventually generate revenue, but there could be a lag of multiple years. Our point here, however, is simply that Austal USA's scale of operations, while having grown considerably, remains sizably below NNS's, whether measured in dollar-value terms or employment levels.

Figure 4.2
The Mobile, Alabama, Region

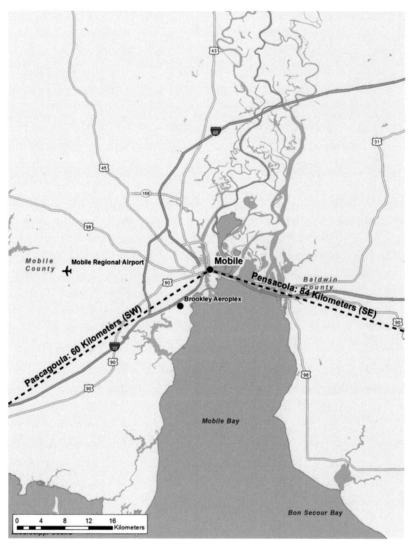

SOURCE: Map created using ArcGIS® software by Esri, with TeleAtlas as the data source. Copyright Esri.

RAND RR1036-4.2

Figure 4.3
The Immediate Vicinity of Austal USA Shipbuilding

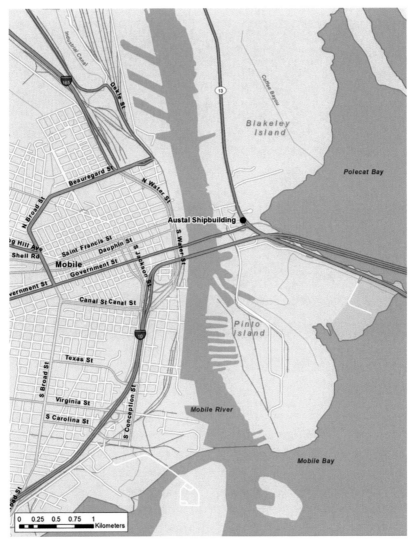

SOURCE: Map created using ArcGIS® software by Esri, with TeleAtlas as the data source. Copyright Esri.

RAND RR1036-4.3

Figure 4.4
A View of Austal Shipbuilding from Downtown Mobile

SOURCE: Edward G. Keating, 2014.
NOTE: LCS 6, *Jackson*, is on the left, and LCS 8, *Montgomery*, is on the right.
RAND *RR1036-4.4*

ing industrial area separated by water from other economic activity and housing.[4] Such an arrangement appears to be preferable to NNS, which abuts the city of Newport News. It is difficult for a retail business that requires access for customers in automobiles to operate adjacent to a shipbuilder, given the shipbuilder's surge of traffic in and out at shift changes. Likewise, it could be unpleasant, due to noise, odors, and traffic, to live next to such an industrial facility. Separation by water mitigates these issues.

---

[4] Ingalls's water separation from the city of Pascagoula is relatively new. The original Ingalls shipyard on the east bank of the Pascagoula River was adjacent to the city and its housing. Today, Ingalls's operations are solely on the west bank of the river, separated from inhabited areas.

Figure 4.5
Downtown Mobile, Alabama

SOURCE: Edward G. Keating, 2014.
RAND *RR1036-4.5*

The Economic Consequences of Austal USA Shipbuilding

Chang (2013) presents data on where Austal USA employees live (see Table 4.1). Not surprisingly, most live in Mobile County or Baldwin County, east across the bay. But there are also a sizable number from the state of Mississippi (Pascagoula, Mississippi, is roughly a 45-minute drive to the southwest of Mobile), as well as some workers who live in the panhandle region of the state of Florida (Pensacola, Florida, is roughly an hour drive to the southeast of Mobile).

Reflecting the fact that shipbuilder jobs are both unique and relatively well paying, we have consistently, across all the shipbuilders we

Table 4.1
Where Austal USA Employees Lived in 2012

Residence	Percentage
Mobile County	60.7
Baldwin County	25.4
Elsewhere in Alabama	2.3
Mississippi	7.9
Florida	3.0
Other	0.7

SOURCE: Chang, 2013.

have examined, found a willingness on the part of shipbuilder employees to undertake sizable driving commutes.[5]

In light of its ramp-up in hiring, Austal USA has considerable recent experience with accession pathways. (A billboard on Interstate 10 notes that Austal is hiring.) While Austal USA has hired some workers with prior shipbuilding experience (e.g., from Ingalls in Pascagoula), even those with prior experience needed considerable training, since Austal ships are aluminum and require aluminum-welding rather than steel-welding expertise. Indeed, most new Austal hires, we were told, lacked any shipbuilding experience. Many lacked any experience in an industrial setting.

To obtain required workforce training for production employees, Austal USA has relied on the State of Alabama–funded Maritime Training Center, shown in Figure 4.6. This facility is located immediately adjacent to the Austal property.

---

[5]   The Austal data in Chang (2013) also provided us with an opportunity to assess the applicability of the U.S. Census Bureau approach in Table 3.1 (Chapter Three) to shipbuilder workers. According to 2011 U.S. Census Bureau data, 71.8 percent of private sector employees in the city of Mobile lived in Mobile County, with 11.9 percent living in Baldwin County, 11.9 percent living elsewhere in Alabama, 2.1 percent in Mississippi, and 1.3 percent in Florida. Shipbuilder employees are more geographically dispersed than the average private sector worker.

Figure 4.6
Alabama Industrial Development Training Maritime Training Center

SOURCE: Edward G. Keating, 2014.
RAND RR1036-4.6

The State of Alabama provides an Austal-designed six-week pre-employment training course to qualified individuals who aspire to work for Austal USA. We were told that Austal USA has ultimately chosen to hire about 60 percent of these trainees. One Austal expert described this state-funded training period as "basically a six-week job interview." Additionally, Austal uses the Maritime Training Center to provide posthiring training and certifications to its new hires.

Perhaps reflecting Austal USA's newness in the local labor market, Austal has experienced considerable annual turnover. We were told that Austal USA has averaged 20-percent annual attrition (which it hopes to reduce to about 15 percent).

Between the roughly 40 percent of state-trained individuals at the Maritime Training Center who are not hired by Austal and individuals who leave Austal employment, Austal has had the effect of considerably altering the workforce-skill profile in the greater Mobile area beyond its current employees.

There is some evidence of an "Austal effect" in Mobile-area economic data. Figure 4.7 shows U.S. Department of Labor, Bureau of Labor Statistics, data on the change in Mobile manufacturing employment since September 2009. We also inserted a line based on the fact that Austal USA told us that its net employment has grown by about 3,400 workers (from about 800 to about 4,200) since late 2009. Austal's employment growth appears to represent a considerable majority of Mobile manufacturing employment growth since late 2009. However, the fact that Mobile's manufacturing growth since late 2009 has exceeded Austal's growth (the red series over the blue line) suggests that Austal's employment growth has not displaced other manufacturing in the region.

Since 2004, Mobile's unemployment rate has roughly tracked the U.S. unemployment rate, as shown in Figure 4.8.

Figure 4.7
Change in Mobile Manufacturing and Austal USA Employment Relative to September 2009

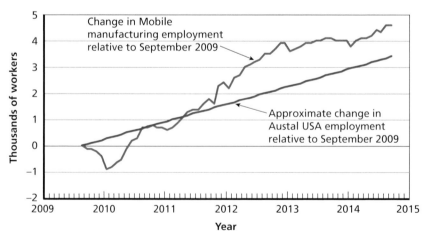

SOURCE: U.S. Department of Labor, Bureau of Labor Statistics, n.d.; and interviews with Austal USA representatives.
RAND RR1036-4.7

Figure 4.8
U.S. and Mobile Unemployment Rates

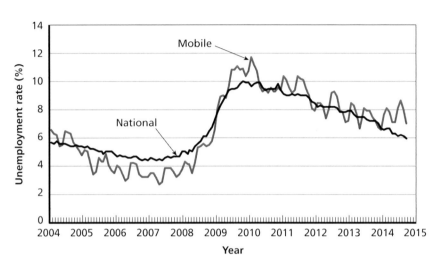

SOURCE: U.S. Department of Labor, Bureau of Labor Statistics, n.d.
RAND RR1036-4.8

Indeed, as shown in Figure 4.9, Mobile's unemployment rate was somewhat higher than the U.S. unemployment rate in 2014.

As of November 2014, Airbus was constructing an A320 aircraft final assembly line in Mobile at the Brookley Aeroplex, about 10 kilometers to the southwest of the Austal facility. This assembly line is scheduled to open in 2015 and will employ around 1,000 people (see Dugan, 2013).

Austal USA has made purchases from local suppliers, but Austal USA's presence has not (at least yet) caused the development of a cluster of nearby suppliers.

The regional economic development experts we interviewed contrasted Austal's effects with those of automobile manufacturers (Mercedes-Benz, Honda, Hyundai) in central Alabama (see Amazing Alabama, 2014). The presence of automobile manufacturers in Alabama, we were told, has caused their various suppliers to build plants near the manufacturers' facilities. A straightforward explanation for this phenomenon in the automobile industry is volume: A car manufacturer building hundreds of thousands of vehicles per year

Figure 4.9
Mobile Unemployment Rate Relative to U.S. Unemployment Rate

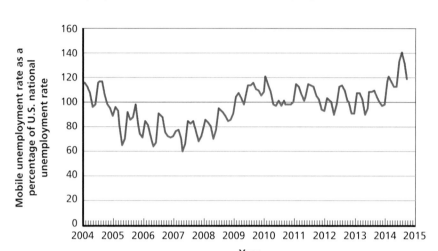

SOURCE: U.S. Department of Labor, Bureau of Labor Statistics, n.d.
RAND RR1036-4.9

needs a commensurable number of seats, headlights, doors, and other parts from suppliers. It is highly advantageous for those suppliers to be located proximate to the automobile manufacturing plant.

By contrast, Austal USA produces a few ships per year. Lacking numerical volume, there is no huge logistical advantage for a supplier to be located near the shipbuilder. Chang's analysis confirms that a large majority (in dollar-value terms) of Austal USA's purchases are from suppliers located outside Alabama (2013).

Insights for Shipbuilding in Australia

Table 1.1 posed questions pertaining to shipbuilding in Australia. The Austal USA case study provides responses to some of those questions, as presented in Table 4.2.

A very different, and much more extensive, supplier clustering effect is discussed in the next chapter's case study, the Gripen program in Sweden.

Table 4.2
Insights for Australian Shipbuilding from the Austal USA Case Study

Opportunity Cost/Displacement	Insights from Austal USA Case Study
What would individuals employed by Austal USA be doing otherwise?	Austal USA has hired many workers without previous experience in a manufacturing setting. These workers have increased their skill levels and, hence, economic opportunities by dint of experience at Austal USA.
To what extent has Austal USA generated favorable spin-offs and spillovers?	While no cluster of suppliers has emerged proximate to Austal USA, the State of Alabama and Austal USA have trained a number of workers who no longer work (and, in some cases, never worked) for Austal USA. These skill-augmented individuals have provided benefits to other industrial employers in the region, we were told.

CHAPTER FIVE

The Gripen Case Study

This chapter examines the issue of spillovers in a different industrial sector, providing an overview of Sweden's JAS-39 Gripen fighter program from Saab Aeronautics. This program has been lauded for successfully delivering an advanced, but affordable, fighter aircraft while also producing a significant economic multiplier within the local and national economies. The possibility of transferring lessons from this apparently successful aerospace project has garnered attention in discussions of Australia's naval shipbuilding industry (e.g., Roos, 2014, Economic Development Board South Australia, 2014).

Limitations with available data and uncertainty with opportunity costs and indirect spillovers make it difficult to quantify the exact value of Gripen to Sweden's economy. However, our on-site interviews made it clear that the Gripen program brought employment to the city of Linkoping in central Sweden, helped anchor a wider aerospace cluster around Saab in the region, and contributed to technological innovation both within major established companies and new firms. Significantly, by focusing on a research and development (R&D) and systems integrator role, Saab was seen to act as a "private technical university," providing knowledge and talent spillovers to the Swedish economy while diffusing a significant portion of the blue-collar manufacturing work across a global network of suppliers (Eliasson, 2010).

Drawing on expert interviews and a literature review, this chapter provides a description of the Gripen program's success, as well as Saab's place in the wider commercialization ecosystem of government, academia, and business in Linkoping. This chapter then informs the discussion in the final chapter of this report as to the lessons that might be applied to the Australian naval shipbuilding context.

Gripen Program Background

The decision to produce a Swedish-made fighter aircraft in the early 1980s was strongly influenced by the country's policy of armed neutrality during the Cold War, by political considerations, and by a particular desire to raise employment in scarcely populated areas (Brandstrom, 2003). The aircraft was produced by the Saab-led IG JAS consortium in Linkoping in central Sweden, about 170 kilometers to the southwest of Stockholm, the nation's political capital and economic center (see Figure 5.1). The program has also been argued to have produced economic spillover throughout Sweden. The project combines elements of the mature NNS example and the more recent growth seen with Austal USA in Mobile. Saab has been present in Linkoping since before the Second World War but never previously pursued a development project of the complexity of the JAS-39 Gripen.

While Saab's site in Linkoping provided the lead on product design, R&D, and final assembly, many systems and subsystems were contracted to suppliers elsewhere in Sweden or overseas, spreading economic spillovers beyond the immediate region. As Carlsson (2010, p. 27) notes, this Saab experience is typical of the aerospace sector, where major aircraft developers are "essentially system integrators[,] . . . [while] manufacturing is instead outsourced to various suppliers in the value chain." The portfolio of suppliers has changed with each Gripen variant and with the differing industrial obligations of the offset agreements signed with foreign customers for the aircraft. For instance, whereas around 35 percent of the Gripen has been produced in the United Kingdom (UK Defence Committee, 2012), future production of the Brazilian Gripen NG is expected to see up to 80 percent of the aerostructure manufactured in Brazil (Stevenson, 2014), including final assembly of 15 of the 36 aircraft ordered (Thisdell, 2014). Major subsystems have been produced in other countries, including France, Germany, South Africa, and the United States.

Figure 5.1
The Linkoping, Sweden, Region

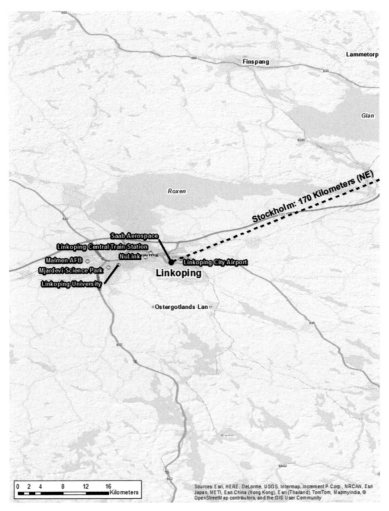

SOURCES: Map created using ArcGIS® software by Esri. Esri, HERE, DeLorme, USGS, Intermap, increment P Corp., NRCAN, Esri Japan, METI, Esri China (Hong Kong), Esri (Thailand), TomTom, MapmyIndia, copyright OpenStreetMap contributors, and the GIS User Community.
NOTE: AFB = Air Force Base.
RAND RR1036-5.1

In 1980, a "stretching" requirement was placed on industry by the Swedish government to produce a lightweight and inexpensive fourth-generation fighter aircraft to replace the country's Viggen and Draken aircraft fleets.[1] The first test flight of a Gripen took place in 1988, but a number of setbacks initially affected the program, including fatal accidents that resulted in urgent revisions to the aircraft's fly-by-wire software. The first deliveries to the Swedish Air Force occurred in 1993; the first-unit F7 wing, based at Satenas, declared initial operating capability in September 1997. While the Swedish government originally ordered 300 aircraft, it ultimately purchased 204, in the post–Cold War climate, with deliveries ending in 2005. A small number of these aircraft flew combat reconnaissance missions over Libya in 2011 as part of Sweden's contribution to the international enforcement of a no-fly zone (DefenceWeb, 2011).

Saab has subsequently worked to upgrade Gripen avionics and weaponry, producing a number of improved single- and two-seat variants for both domestic and foreign buyers. These aircraft have provided the basis for an even more advanced, network-centric "next-generation Gripen," as well as the ongoing development of a carrier-borne "sea Gripen" and an optionally manned variant (IHS Janes, 2014).

In addition to sales to the Swedish military and Brazil, Saab has exported the aircraft to the Czech Republic, Hungary, South Africa, and Thailand, as well as the United Kingdom's Empire Test Pilots' School. The company is also in discussions with the Malaysian government and has received expressions of interest from Argentina and Botswana. Lennart Sindahl, head of Saab Aeronautics, has set a target of worldwide sales between 300 and 450 Gripen C/D/E aircraft over the next 20 years—equal to 10 percent of the accessible global market (IHS Janes, 2014). While, in the early 2000s, Saab derived around 70 percent of its revenues from the Swedish government and 30 percent from overseas, that ratio has been reversed in recent years (Morrison, 2012).

In 1996, the Swedish government budgeted around 60 billion Swedish krona (SEK) for the program, which translated into about

---

[1]   This paragraph draws on material from IHS Janes (2014).

100 billion SEK in 2007 terms.[2] The program experienced around 9.3 billion SEK in cost growth to 2002 (IHS Janes, 2014), with program costs totaling 122 billion SEK by 2007 (see Table 5.1). Further development funding has subsequently been used to develop the Next Generation Gripen, with academic analyses of the spillover effects from Gripen stopping at this transition point in 2007 (Eliasson, 2010).

Spillovers

While the experts we interviewed for this study stressed that direct employment is not the most significant metric indicating a project's spillover success, job creation was an important political consideration in the original decision to proceed with Gripen. The initial 1982 agreement between the Swedish government and IG JAS, a Saab-led industrial consortium, set a target for the creation of 800 jobs in a region with high unemployment (Skons and Wetterqvist, 1994, p. 229). By 1987, the Gripen program had generated an estimated 1,200 new jobs

Table 5.1
Gripen Program Costs, 1982 to 2007

Period and Category	Costs (billions of 2007 SEK)
1982 through 1992	
R&D	32.4
Manufacturing	5.6
1992 through 2007	
R&D	44.6
Manufacturing	39.6

SOURCE: Eliasson, 2010, p. 257.

---

[2]   As of December 2014, the SEK was trading at about AUD 0.15 and USD 0.14. Hence, Table 5.1's values sum to about AUD 18 billion. However, the translation is imprecise because the SEK–Australian dollar exchange rate has varied sizably over time.

(Statens offenstliga utredningar, 1993, p. 119; Ahlgren et al., 1998). As Brandstrom (2003, p. 18) notes:

> As the largest industrial project undertaken in Sweden, the JAS project directly supported the development of civil aviation and was calculated to indirectly "spill-over" into other vital parts of Swedish industry. The spillover effects included enhancing productivity and production in other parts of the economy. Employment prospered in several parts of the aircraft industry and hence some of the new employment opportunities were located in the more sparsely populated areas.

Saab currently has about 700 industrial workers tasked on the Gripen in Sweden, though there are hopes to create 1,000 new jobs in Linkoping over the next 20 years to manage exports of Gripen E and other next generation models (SvD Naringsliv, 2014). Stefan Folster, chief economist at the Confederation of Swedish Enterprise, calculates that 3,000 jobs have been created in Sweden as a direct result of Gripen (Saab, n.d.). The wider Linkoping "aerospace cluster" currently employs around 18,000 workers, one-third of the city's labor force (Flyghuvudstad.se, n.d.). Eliasson (2010, p. 5) stressed:

> A small industrial country such as Sweden with nine million inhabitants has been capable of developing one of the world's most advanced combat aircraft systems without draining its industry of engineering resources. . . . The spillovers from the Gripen project have been so large and have represented such a large resource input in production that neither society nor industry suffered. On the contrary, both . . . benefitted[,] . . . perhaps several times over.

Interviewees noted that the intention of the Swedish government from the outset was to use the program to incentivize wider economic growth—bringing intangible benefits in terms of knowledge spillovers, talent development, and new industrial partnerships, not just the more readily apparent creation of jobs. As Berkok, Penney, and Skogstad (2012, p. 4) note:

Sweden's aerospace industry—and the Gripen program in particular—has acted as the country's main driver in defence systems development and innovation. This particular program has always relied on the smart and cooperative procurement approach with a long-term partnership in mind where the purchaser participates in development. Accordingly, Swedish defence industrial policy is geared toward supporting domestic developers through the use of subsidies to [small and medium-sized enterprises] and offset policies with elements of technology transfers and cooperation in R&D.

In addition to new civilian production by existing companies, technology spillovers from the Gripen program fed into a number of technology companies being spun out. Both expert interviewees and academic literature attribute a varying portion of the value created in a number of new firms to the Gripen project (Eliasson, 2010). The complex engineering problems posed by a high-technology aircraft acted as a spur to innovation, along with the exchange (both formal and informal) of knowledge and talent between Saab and its industrial partners in Sweden and abroad. Developing Gripen necessitated a number of innovations in design methods, project management, machining techniques, composite materials, and advanced electronics that have proved to have commercial applications, in some cases far outside aviation (see Table 5.2).

Table 5.2
Firms That Emanated from the Gripen Program

Firm	Business Description
MX Composites	Composite materials for engine components
Nobel Biocare	Advanced solutions for tooth implants (derived from Gripen materials)
Biosensor Applications	Artificial "nose" for detecting drugs or explosives
Combitech Traffic Systems	Aviation traffic management and software
SMM Medical	Treatments for cardiovascular diseases (derived from flight suits)

SOURCE: Eliasson, 2010.

In addition to spin-offs, interviewees and the academic literature emphasize that major Swedish companies captured a significant value of innovation internally, developing a range of new product lines and production methods as a result of lessons learned from the Gripen program. In the case of Volvo, for instance, the firm's role in developing engines for the IG JAS consortium is seen as highly significant to its subsequent move into the civilian aviation engine market. Eliasson (2010) and expert interviewees assert that Gripen was particularly important to Ericsson, whose global success in mobile telecommunications came after the development of significant expertise in military communications for the Gripen.

In its systems integrator role, Saab itself captured a number of spillover technologies, developing a range of complex design, production, and computer modeling techniques that have subsequently been applied to other products. In addition to producing a range of new product lines, such as visors for firefighters (derived from the Gripen canopy), new engines, and control software (Eliasson, 2010), Saab's emphasis on lightweight materials for Gripen enabled the firm to reposition itself as a tier-1 supplier of lightweight materials and components for commercial aircraft for Airbus and Boeing.

Interviewees told us that Saab's involvement in offset deals with foreign Gripen customers has also helped it and the wider Linkoping cluster establish strategic partnerships with firms and governments overseas. The Mjardevi Science Park in Linkoping, for instance, was involved in establishing similar innovation campuses in South Africa as part of Saab's offset obligations with that country. Swedish companies, such as Electrolux, have established factories in the Czech Republic and Hungary (Saab, n.d.), while Saab is hosting 100 Brazilian engineers and their families in Linkoping as part of a sale of 36 Gripen NG aircraft to that country (Corren.se, 2014).

The Gripen Program's Economic Multiplier

Perhaps not surprisingly, in light of the program's reported successful spillovers, the Gripen program appears to have had a larger (more favorable) economic multiplier than most of the cases discussed in Chapter Two.

Eliasson (2010) estimates that the total Swedish government R&D investment in the program was about 130 billion SEK in 2007 terms.[3] That R&D investment then resulted in about 340 billion SEK in what he terms "net social value creation"—that is, value on top of the delivery value of the aircraft itself—an estimate Eliasson produced through the aggregation of estimates of the value added in a large sample of Swedish companies and spin-offs. Using the multiplier formulation employed by the literature discussed in Chapter Two, this implies that the program had an economic multiplier of about 3.6:

$$3.6 = \left(\frac{340 + 130}{130} \right)$$

Interviewees emphasized the challenges and uncertainties latent in a multiplier estimate of this sort. The commercial benefits of knowledge spillovers from a project may only become apparent years later. It is also difficult to distinguish the influence of Gripen from a range of other necessary factors that contribute to commercial or technological breakthroughs.

Gripen Program Discussion

The Gripen program's apparent economic success provokes two key questions. First, what actions by the Swedish government and characteristics of the program can be credited with this success? Second, to what extent is the Gripen program an analogy that is relevant to Australian shipbuilding? In the remainder of this chapter, we discuss the first question—i.e., how and why the Gripen program seems to have been so successful. The discussion in Chapter Six will tackle the second question, the applicability of the analogy (for all three of our case studies).

---

[3] This 130 billion SEK total is somewhat higher than the sum of the totals in Table 5.1; this is because Eliasson (2010) additionally included a 4-percent real interest rate into this calculation.

We divided explanations for Gripen's success into three categories that we will discuss in turn:

- the importance of advanced R&D as a high-intensity spillover multiplier
- the need for a wider commercialization ecosystem to capture spillovers
- the central role of government as an "advanced customer," both in setting stretching requirements for the Gripen and in fostering the wider infrastructure to harness the resultant spillovers of knowledge and talent.

The Importance of Advanced R&D

Subject-matter expert interviewees and the academic literature both place significant emphasis on the product-development phase as the key driver of spillovers from the Gripen program. The manufacturing work that came subsequently had a much lower multiplier (Eliasson, 2011, p. 258). This finding is consistent with Moretti (2002), who argued for the importance of complex design problems in generating knowledge and innovation.

As a company, Saab invests some 20 percent of its revenue in R&D. Its workforce demographics reflect that emphasis, with more than 75 percent of Saab staff classified by the firm as white collar at the height of the Gripen program and about 20 percent of employees possessing postgraduate degrees (Saab, 2004, p. 29). Interviewees told us that the challenge of making Gripen lightweight and affordable necessitated a particular focus on research in materials science and computer modeling that has benefited Saab's emergence as a supplier of composite components to Airbus and Boeing.

Gripen's high proportion of white-collar employees is very different from NNS and Austal USA. Both shipbuilders have labor forces dominated by blue-collar or production workers. This workforce mix difference may be a broader point of contrast between the aerospace and shipbuilding industries.

A significant portion of Gripen manufacturing work has been diffused across Saab's supply chain, both throughout Sweden and internationally. Saab's strategy of making technology-transfer and offset deals regarding employment a central part of its offer to foreign Gripen customers has compounded this trend, with Saab committing to job-creation targets in many customer countries (IHS Janes, 2014). In Hungary, for instance, Saab claimed to deliver more than 10,000 new jobs through a combination of greenfield investments and new export projects (Saab, n.d.).

The Need for a Wider Commercialization Ecosystem

Another key lesson of the Gripen program is the significance placed by both the literature and expert interviewees on establishing a fertile commercialization ecosystem in which spillovers from a major industrial project can be nurtured and captured for the local and national economies. Eliasson (2010, 2011) emphasizes the need for the local economy and government to have sufficient competence to commercialize the opportunities presented by knowledge spillovers, noting that Saab acted as a "private technical university" in supplying innovations to the Swedish economy without being able to capture more than a small portion of the wider value.

Linkoping's aerospace cluster has been the subject of a number of studies that underscore the importance of these local commercialization networks. Klofsten, Jones-Evans, and Scharberg (1999) refer to a "triple helix" model in Linkoping, charting a series of overlapping, mutually reinforcing partnerships between industry (including, but not limited to, Saab), Linkoping University (LiU) and local government. It is important to stress that this triple helix represents the end-point of a long-standing, iterative process, stretching back to Saab's decision to establish aviation industry in Linkoping in the 1940s. Successive Saab aircraft programs, such as the Draken and Viggen, helped anchor—and were in turn supported by—the development of the wider technical and commercial competence of the local area. Of particular significance was the establishment of LiU in the 1960s and 1970s (following lobbying from Saab), with interviewees noting the university's vocational approach, support for entrepreneurship, and focus on then-nascent fields, such as computer science. The key has been developing an extensive knowledge base that can be utilized by other actors when attempting to capture the economic value of technology spillovers for the local economy. Such bodies as LiU not only helped provide Saab with a pool of skilled labor but also created a critical mass of support across local industry, academe, and government for fostering spillovers through joint investment in the necessary physical, regulatory and financial infrastructure. As Klofsten, Jones-Evans, and Scharberg (1999, p. 125) note, "The development that has taken place over the last thirty years can be linked to a spiral where success begets success to foster a positive entrepreneurial climate."

In addition to more-informal channels of circulating knowledge and talent between local entities, Saab maintains formal ties, especially with LiU, with funding for teaching, research, industrial PhDs, and adjunct professors (Klofsten and Jones-Evans, 1996). A number of local bodies exist to encourage spin-offs and new technology start-ups, including the Technology Bridge Foundation, the Foundation for Small Business Development, and the Centre for Innovation and Entrepreneurship. The Mjardevi Science Park's LiU Entrepreneurship and Development (LEAD) incubator, with funding and support from LiU, has won awards as one of Europe's leading sites for developing new technology-based businesses.

Internally, Saab maintains its own organization, Saab Ventures, to promote spin-off technologies from Gripen and its other projects. Saab Ventures has invested USD 35 million in developing Saab spin-offs with three companies—C3 Technologies, A2 Acoustics, and Combitech—with combined revenue of around USD 120 million in 2012 (Konda, 2012). Successful spin-off companies now work in diverse fields, such as soundproofing, video surveillance, and medical equipment (NuLink, n.d.).

The Central Role of Government

Eliasson, the program's prominent advocate, credited the Swedish government for its role as what he terms "an advanced customer." The Swedish government had the dual role of both purchasing the military platform and trying to maximize favorable spillovers from it. Eliasson (2010, p. 166), notes:

First, the government should encourage the development of a competent local commercialization industry. Second, however, the government also carries the responsibility to act as a substitute customer of privately demanded public goods and services that will not be supplied in the market without that mediation.

He argues that investment decisions should be made in such a way as to move beyond a traditional customer-supplier relationship and take into account not only immediate acquisition goals (e.g., to acquire a high-quality platform at low cost) but also the complex and cascading effects on wider industry. This requires market intelligence about existing linkages, or lack thereof, between the triple helix of government, academia, and industry, in addition to a detailed, realistic understanding of the commercialization capacity of the local region hosting the project.

Interviewees noted the willingness of the Swedish government to promote the commercialization of advanced R&D on the Gripen program. This approach offers synergies with the country's avowed aim to maintain core sovereign capabilities in defense production as part of Sweden's neutrality policy. Interviewees also stressed that Sweden learned from unfavorable experiences in the 1970s, when the Swedish government lost large amounts of money attempting to support its civilian shipbuilding industry that had become uncompetitive relative to East Asian producers.

As one interviewee noted, the Swedish government has taken an active and intelligent role as the primary customer for Gripen. The government understood the technical requirements and industrial challenges well enough to push Saab to a point where it was forced to innovate and develop R&D spillovers. At the same time, the Swedish government did not push so far that Saab was asked to provide systems or components that were beyond its competence levels. Instead, such systems and components were subcontracted to foreign suppliers. The project emphasized Swedish industry's strengths in advanced R&D and complex systems integration, promoting sovereign capabilities, but not at the expense of affordability.

Insights for Shipbuilding in Australia

Table 1.1 posed questions pertaining to shipbuilding in Australia. The Gripen case study provides response to some of those questions, as presented in Table 5.3.

Table 5.3
Insights for Australian Shipbuilding from Gripen Case Study

Opportunity Cost/Displacement	Insights from Gripen Case Study
To what extent has the Gripen project generated favorable spin-offs and spillovers?	Gripen generated very significant numbers of spin-offs and spillovers across a number of realms, including nondefense.
Are there ways to structure a hybridization to maximize the extent of favorable spin-offs/spillovers?	The Gripen program largely employed white-collar workers with skills in such areas as system integration, complex design, computer modeling, composite materials, and advanced electronics. A hybridization that concentrates on high-end skills might generate greater levels of spin-offs/spillovers.

CHAPTER SIX

Discussion

Table 6.1 reprises Table 1.1, enumerating Australian shipbuilding alternatives and questions raised against different economic consequence rubrics.

In terms of what individuals employed in Australian shipbuilding would be doing otherwise, key issues are the state of the Australian (and the shipbuilder's regional) economy and the degree of difficulty these workers would have finding commensurate alternative employment.

Table 6.1
Australian Shipbuilding Alternatives and Questions About Economic-Consequence Rubrics, Reprisal

Approach	Opportunity Cost/Displacement	Spin-Offs/Spillovers
Indigenous production	What would individuals employed in Australian shipbuilding be doing otherwise?	To what extent would shipbuilding in Australia generate favorable spin-offs and spillovers?
Purchase abroad	Assuming that foreign-built ships cost less, what would the Australian government and/or taxpayers do with cost savings?	Could the Australian government or Australian taxpayers invest cost savings in a realm that generates more-favorable spin-offs and spillovers?
Hybridization (buy some ships or some parts of ships abroad)	Which skill sets would be required in Australia under the hybridization, and what would be the opportunity cost and displacement associated with these workers?	Are there ways to structure the hybridization to maximize the extent of favorable spin-offs and spillovers?

Our examination of shipbuilders in the United States found slack economies in both regions and considerable rigidity in workers' abilities to find commensurate employment. Workers employed in shipbuilding appear to be quite geographically immobile (though willing to incur sizable driving commutes). Both NNS and Austal USA are able to attract many job applicants, suggesting that these workers do not have alternative employment options as desirable as working at the shipbuilders. (For both NNS and Austal USA, the central challenge has not been in the number of applicants but in their skill sets.) Several experts noted a tendency for laid-off shipbuilder workers to have prolonged periods of unemployment or underemployment, awaiting recall to the shipbuilder. The shipbuilders have not displaced high-value activities for many of their workers.

Regarding the extent to which shipbuilding in Australia would generate favorable spin-offs and spillovers, the U.S. examples are not optimistic. For example, NNS appears to have generated relatively few spillovers. Indeed, the entire Hampton Roads region has been critiqued for a dearth of entrepreneurial activity. Likewise, no cluster of suppliers has yet emerged around Austal USA. Eliasson (2010, Table 5.2) has no analog in shipbuilding in the United States, to our knowledge.

One inhibitor of clustering or spillovers is the limited production volume in shipbuilding. While an automotive supplier has strong incentive to minimize logistics costs by locating near the manufacturer, the proximity incentive is far less in low-volume shipbuilding.

In contrast to Gripen's success in the aviation industry, several experts we interviewed suggested that shipbuilding tends to have fewer creative spillovers. While the aviation industry has developed such technologies as composite materials and advanced adhesives, analogs in shipbuilding are hard to identify (though one expert noted considerable progress in paint technology in shipbuilding). The Gripen analogy appears to be overly optimistic as to the magnitude and nature of spin-offs and spillovers that might be expected from naval shipbuilding in Australia.

If the Australian government wished to pursue a hybridization approach, from a spin-offs and spillovers perspective, it would be desirable to hold onto the "Gripen-like" aspects of shipbuilding. We were told that outfitting and systems integration are the sorts of more high technology– and R&D-oriented activities that have provided so many favorable spillovers in the Gripen context. Blue-collar activities appear to be less prone to generate favorable spillovers.

A high technology–oriented hybridization would not be without concern, however. In particular, if the goal is to hybridize in a manner that minimizes opportunity cost and displacement, it is exactly the blue-collar workers—who were but a minor portion of the Gripen workforce—who appear to benefit most directly from ship production at NNS and Austal USA. One of our subject-matter experts suggested that NNS's white-collar workers would probably live and work elsewhere in the United States in other industries—i.e., the opportunity cost of shipbuilding employment is considerable. Instead, it is often the blue-collar or production workers who have the greater incremental gain and lower opportunity cost from employment at the shipbuilder.

Table 6.1's opportunity cost and displacement and spin-offs and spillovers columns may therefore trade off against one another. Indeed, NNS and Austal USA rate much better against opportunity cost and displacement metrics; the Gripen program rates better against spin-offs and spillovers metrics. Indigenous production of ships in Australia cannot be expected to have both low opportunity cost and displacement and high levels of favorable spillovers.

Abbreviations

AUD	Australian dollars
IMPLAN	Impact Analysis for Planning
JHSV	Joint High Speed Vessel
LCS	Littoral Combat Ship
LiU	Linkoping University
MAR	Marshall-Arrow-Romer
MIG	Minnesota IMPLAN Group
NNS	Newport News Shipbuilding
NNSY	Norfolk Naval Shipyard
R&D	research and development
RCOH	refueling complex overhaul
RIMS	Regional Input-Output Modeling System
SEK	Swedish krona
USD	U.S. dollars

References

Acil Allen Consulting, *Naval Shipbuilding & Through Life Support: Economic Value to Australia, Maintaining Capabilities and Capacity*, Brisbane, Qld., 2013.

Ahlgren, J., L. Christofferson, L. Jansson, and A. Linner, *Faktaboken om Gripen*, 4th ed., Linkoping, Sweden: Industrigruppen JAS AB, 1998.

Amazing Alabama, "Automotive Hub of the South," July 21, 2014. As of December 8, 2014:
http://www.amazingalabama.com/key-industry-targets-automotive.html

Arena, Peter M., Roger R. Stough, and Mark Trice, *The Contribution of Newport News Shipbuilding to the Virginia Economy: 1996 with Forecasts to 2020*, Arlington, Va.: Center for Regional Analysis, George Mason University, 1996.

Austal Limited, *2010: Concise Report*, 2010. As of March 13, 2015:
http://media.corporate-ir.net/media_files/IROL/15/159601/annualRep/AR2010.pdf

————, annual reports, 2009–2014. As of March 31, 2015:
http://investor.austal.com/phoenix.zhtml?c=159601&p=irol-reportsAnnual

Barro, Robert J., and Charles J. Redlick, "Macroeconomic Effects from Government Purchases and Taxes," *The Quarterly Journal of Economics*, Vol. 126, No. 1, 2011, pp. 51–102.

Berkok, Urgurhan, Christopher Penney, and Karl Skogstad, "Defense Industrial Policy Approaches and Instruments," unpublished manuscript, July 2012. As of November 1, 2014:
http://www.econ.queensu.ca/files/other/
Defence%20Industrial%20Policy%20Approaches%20and%20Instruments.pdf

Bozick, Tara, "ODU Report: Decline in Defense Spending Stalled Hampton Roads Jobs Recovery," *Daily Press*, October 7, 2014. As of October 31, 2014:
http://www.dailypress.com/business/tidewater/dp-odu-report-hampton-roads-economy-20141007-story.html

Brandstrom, Annika, *Coping with a Credibility Crisis: The Stockholm JAS Fighter Crash of 1993*, Stockholm, Sweden: Swedish National Defence College, 2003.

Carlino, Gerald A., "Knowledge Spillovers: Cities' Roles in the New Economy," *Business Review*, 4th Q., 2001, pp. 17–26. As of December 26, 2014: http://www.philadelphiafed.org/research-and-data/publications/business-review/2001/q4/brq401gc.pdf

Carlsson, Bo, "Creation and Dissemination of Knowledge in High-Tech Industry Clusters," unpublished manuscript, April 21, 2010. As of December 1, 2014: http://www.wiwi.uni-jena.de/eic/files/SS%2011%20JERS%20Carlsson.pdf

Chang, Semoon, "Austal USA: Economic Impact on the Coastal Counties of Alabama," Gulf Coast Center for Impact Studies, April 24, 2013.

Cohen, Lauren, Joshua Coval, and Christopher Malloy, "Do Powerful Politicians Cause Corporate Downsizing?" draft, Harvard Business School and National Bureau of Economic Research, October 7, 2011.

Cooper, Elizabeth, "Revitalization of Newport News Neighborhood Shows City's Agenda," *Virginia Business*, September 27, 2013. As of November 2, 2014: http://www.virginiabusiness.com/news/article/revitalization-of-newport-news-neighborhood-shows-citys-agenda

Corren.se, "Brasilianska ingenjörer till Saab" ["Brazilian Engineers at Saab"], October 9, 2014. As of December 2, 2014: http://www.corren.se/ekonomi/brasilianska-ingenjorer-till-saab-7410763.aspx

DefenceWeb, "Swedish Gripens Deliver 37% of Libyan Reconnaissance Reports," August 8, 2011. As of December 1, 2014: http://www.defenceweb.co.za/index.php?option=com_content&view=article&id=17886:swedish-gripens-deliver-37-of-libyan-reconnaissance-reports&catid=35:Aerospace&Itemid=107

Deloitte Access Economics, *The Economic Impact of Major Defence Projects*, December 2014.

Dugan, Kelli, "Airbus Hiring on Track for A320 Final Assembly Line, Efficiency Driving Design," *AL.com*, October 15, 2013. As of December 8, 2014: http://blog.al.com/live/2013/10/airbus_hiring_on_track_for_a32.html

Economic Development Board South Australia, "Economic Analysis of Australia's Future Submarine Program," October 2014.

Eliasson, Gunnar, *Advanced Public Procurement as Industrial Policy*, New York: Springer, 2010.

———, "The Commercialising of Spillovers: A Case Study of Swedish Aircraft Industry," in Andreas Pyka, Derengowski Fonseca, Maria da Graca, eds., *Catching Up, Spillovers and Innovation Networks in a Schumpeterian Perspective*, New York: Springer, 2011, pp. 147–170.

Feldman, Maryann P., and David B. Audretsch, "Innovation in Cities: Science-Based Diversity, Specialization, and Localized Competition," *European Economic Review*, Vol. 43, No. 2, February 15, 1999, pp. 409–429. As of January 6, 2015: http://www.sciencedirect.com/science/article/pii/S0014292198000476

Fisher, Jonas D. M., and Ryan Peters, "Using Stock Returns to Identify Government Spending Shocks," *The Economic Journal*, Vol. 120, May 2010, pp. 414–436.

Flyghuvudstad.se, "About the Aviation Capital," n.d. As of November 1, 2014: http://flyghuvudstaden.se/en/about-flyghuvudstaden/

Glaeser, Edward L., Hedi D. Kallal, Jose A. Schienkman, and Andrei Shleifer, "Growth in Cities," *Journal of Political Economy*, Vol. 100, No. 6, 1992, pp. 1126–1152.

Hall, Robert E., "The Role of Consumption in Economic Fluctuations," in Robert J. Gordon, ed., *The American Business Cycle: Continuity and Change*, Chicago: University of Chicago Press, 1986, pp. 237–266.

———, "By How Much Does GDP Rise If the Government Buys More Output?" Cambridge, Mass.: National Bureau of Economic Research, Working Paper 15496, November 2009.

Haun, Eric, "Bagley Named VP at Newport News Shipbuilding," *MarineLink. com*, November 14, 2014. As of November 16, 2014: http://www.marinelink.com/news/shipbuilding-newport380827.aspx

Herman, Arthur, *Freedom's Forge: How American Business Produced Victory in World War II*, New York: Random House, 2012.

Hosek, James, Aviva Litovitz, and Adam C. Resnick, *How Much Does Military Spending Add to Hawaii's Economy?* Santa Monica, Calif.: RAND Corporation, TR-996-OSD, 2011. As of March 13, 2015: http://www.rand.org/pubs/technical_reports/TR996.html

Huntington Ingalls Industries, "Who We Are: About Newport News Shipbuilding," 2014. As of November 3, 2014: http://nns.huntingtoningalls.com/about/index

IHS Janes, "Saab JAS-39 Gripen," Jane's All the World's Aircraft, August 26, 2014. As of November 1, 2014: https://janes.ihs.com/CustomPages/Janes/DisplayPage.aspx?ShowProductLink= true&DocType=Reference&ItemId=+++1343049&Pubabbrev=JAWA

Ironfield, Denise, *Impact of Major Defence Projects: A Case Study of the ANZAC Ship Project*, Lyneham, ACT: Tasman Asia Pacific Pty Ltd., February 2000.

———, *Impact of Major Defence Projects: A Case Study of the Minehunter Coastal Project*, Belconnen, ACT: Tasman Economics, January 2002.

Jacobs, Jane, *The Economy of Cities*, New York: Vintage, 1969.

Klepper, Steven, "The Origin and Growth of Industry Clusters: The Making of Silicon Valley and Detroit," *Journal of Urban Economics*, Vol. 67, No. 1, January 2010, pp. 15–32. As of January 5, 2015: http://www.sciencedirect.com/science/article/pii/S0094119009000655

Klofsten, Magnus, and Dylan Jones-Evans, "Stimulation of Technology-Based Small Firms—A Case Study of University-Industry Cooperation," *Technovation*, Vol. 16, No. 4, 1996, pp. 187–193.

Klofsten, Magnus, Dylan Jones-Evans, and Carina Schärberg, "Growing the Linköping Technopole—A Longitudinal Study of Triple Helix Development in Sweden," *Journal of Technology Transfer*, Vol. 24, 1999, pp. 125–138.

Koch, James, "The State of the Region: Hampton Roads 2014," briefing slides, Norfolk, Virginia, October 7, 2014.

Konda, Rafael, "Saab Ventures negocia parcerias no país," Instituto de Tecnologia de Sao Caetano Do Sol, November 21, 2012. As of December 2, 2014: http://www.itescs.com.br/saab-ventures-negocia-parcerias-no-pais/

Lessig, Hugh, "Retiring Newport News Shipbuilding Exec Leaves Enduring Legacy," *Daily Press*, December 13, 2014. As of December 17, 2014: http://www.dailypress.com/news/military/dp-nws-evg-shipyard-hunley-retires-20141213-story.html#page=1

———, "Apprentice School: More Than Learning a Trade in Newport News," *Daily Press*, February 8, 2015. As of February 9, 2015: http://www.dailypress.com/news/military/dp-nws-apprentice-school-20150208-story.html

Manski, Charles F., "Removing Deadweight Loss from Economic Discourse on Income Taxation and Public Spending," *VOX*, August 18, 2013. As of January 18, 2015: http://www.voxeu.org/article/removing-deadweight-loss-economic-discourse-income-taxation-and-public-spending

McCabe, Robert, "Naval Shipyard Hiring for 1,500 Jobs in Coming Year," *Norfolk Virginian-Pilot*, December 19, 2014. As of December 19, 2014: http://hamptonroads.com/2014/12/naval-shipyard-hiring-1500-jobs-coming-year

Moretti, Enrico, "Human Capital Spillovers in Manufacturing: Evidence from Plant-Level Production Functions," Cambridge, Mass.: National Bureau of Economic Research, NBER Working Paper No. 9316, 2002.

Morgan, Jonathan Q., "Analyzing the Benefits and Costs of Economic Development Projects," University of North Carolina, Chapel Hill, *Community and Economic Development Bulletin*, No. 7, April 2010.

Morrison, Murdo, "In Focus: Saab Restructures to Better Target Export Markets," *Flight Global*, November 23, 2012. As of December 1, 2014:
http://www.flightglobal.com/news/articles/
in-focus-saab-restructures-to-better-target-export-379256/

Mourshed, Mona, Diana Farrell, and Dominic Barton, *Education to Employment: Designing a System That Works*, New York: McKinsey & Company, January 2013. As of December 24, 2014:
http://mckinseyonsociety.com/downloads/reports/Education/Education-to-Employment_FINAL.pdf

Nakamura, Emi, and Jon Steinsson, "Fiscal Stimulus in a Monetary Union: Evidence from U.S. Regions," working paper, New York: Columbia University, September 2, 2013.

Naval Sea Systems Command, *2012 Command Pocket Guide*, Washington, D.C., 2012. As of November 17, 2014:
http://www.navsea.navy.mil/OnWatch/assets/images/pocket_guide_2012.pdf

Newport News Shipbuilding, "Fact Sheet," n.d. As of November 3, 2014:
http://nns.huntingtoningalls.com/about/NNS-Fact-Sheet.pdf

NNS—*See* Newport News Shipbuilding.

Norfolk Department of Development, "Maritime and Transportation," n.d. As of November 30, 2014:
http://www.norfolkdevelopment.com/
index.php?option=com_content&view=article&id=137&Itemid=69

NuLink, *Linköpings näringsliv [Business Linkoping 2011/2012]*, Linkoping: Sweden, n.d. As of December 2, 2014:
http://iqpager.quid.se/pdf2swf/files/User861507900/UploadedPDF/Lkpg_närliv_svenskA.pdf

Oanda Corporation, "Historical Exchange Rates," web page, n.d. As of November 29, 2014:
http://www.oanda.com/currency/historical-rates/

Owyang, Michael T., Valerie A. Ramey, and Sarah Zubairy, "Are Government Spending Multipliers Greater During Periods of Slack? Evidence from Twentieth-Century Historical Data," *American Economic Review: Papers & Proceedings*, Vol. 103, No. 3, 2013, pp. 129–134.

Porter, Michael E., *Competitive Advantage of Nations*, New York: The Free Press, 1990.

Regional Studies Institute, *The State of the Region: Hampton Roads 2014*, Norfolk, Va.: Old Dominion University, 2014. As of October 31, 2014:
https://www.odu.edu/forecasting/state-region-reports/2014

Roos, Göran, "Future of Australia's Naval Shipbuilding Industry," supplementary submission to the Senate Economics References Committee, Canberra: Senate Printing Unit, Parliament House, October 13, 2014. As of March 31, 2015: http://www.defence.gov.au/Whitepaper/docs/135-GovernmentofSouthAustralia. pdf

Saab, *Gripen—the Face of Success*, Stolkholm, Sweden, n.d. As of November 1, 2014: http://www.saabgroup.com/Global/Documents%20and%20Images/Air/Gripen/ Gripen%20product%20sheet/The_Face_of_Success.pdf

———, *Saab 2003*, Stockholm, Sweden, January 11, 2004. As of December 2, 2014: http://www.saabgroup.com/Global/Documents%20and%20Images/About%20 Saab/Investor%20relations/Financial%20reports/2003_Annual_report_part1.pdf

Schank, John F., Mark V. Arena, Denis Rushworth, John Birkler, and James Chiesa, *Refueling and Complex Overhaul of the USS Nimitz (CVN 68): Lessons for the Future*, Santa Monica, Calif.: RAND Corporation, MR-1632-Navy, 2002. As of November 3, 2014: http://www.rand.org/pubs/monograph_reports/MR1632.html

Skons, Elisabeth, and Fredrik Wetterqvist, "The Gripen and Sweden's Evolving Defence Industrial Policy," in Randall Forsberg, *The Arms Production Dilemma: Contraction and Restraint in the World Combat Aircraft Industry*, Cambridge, Mass.: MIT Press, 1994, pp. 217–238.

Statens offenstliga utredningar, *JAS 39 Gripen—en granskning av JAS-projektet*, Stockholm, Sweden, 1993.

Stevenson, Beth, "Brazilian Air Force Confirms Gripen Acquisition Numbers," *Flight Global*, November 18, 2014. As of December 1, 2014: http://www.flightglobal.com/news/articles/ brazilian-air-force-confirms-gripen-acquisition-numbers-406213/

SvD Naringsliv, "Saab anställer 1000 personer," March 20, 2014. As of November 1, 2014: http://www.svd.se/naringsliv/saab-anstaller-1-000-personer-pa-20-ar_3379010.svd

Thisdell, Dan, "Saab, Brazil Finalise Gripen NG Deal," *Flight Global*, October 27, 2014. As of December 1, 2014: http://www.flightglobal.com/news/articles/ saab-brazil-finalise-gripen-ng-deal-405288/

UK Defence Committee, "Written Evidence from Saab," London: Parliament, April 2012. As of December 1, 2014: http://www.publications.parliament.uk/pa/cm201213/cmselect/cmdfence/9/9vw10. htm

USASpending.gov, website, n.d. As of November 3, 2014: http://www.usaspending.gov/

U.S. Census Bureau, "American FactFinder," March 2014a. As of March 20, 2015:
http://factfinder.census.gov/faces/nav/jsf/pages/community_facts.xhtml

———, "State & County QuickFacts," July 8, 2014b. As of December 8, 2014:
http://quickfacts.census.gov/qfd/states/01/0150000.html

U.S. Census Bureau, Center for Economics Studies, *OnTheMap*, n.d. As of November 10, 2014:
http://onthemap.ces.census.gov/

U.S. Department of Commerce, Bureau of Economic Analysis, "National Data," Section 1, Table 1.1.9, Implicit Price Deflators for Gross Domestic Product, November 25, 2014. As of November 29, 2014:
http://www.bea.gov/iTable/iTable.cfm?ReqID=9&step=1#reqid=9&step=3&isuri=1&903=13

U.S. Department of Defense, "Secretary of Navy Names High Speed Ferries Guam and Puerto Rico," news release, No. 358-12, May 8, 2012. As of December 8, 2014:
http://www.defense.gov/releases/release.aspx?releaseid=15255

U.S. Department of Labor, Bureau of Labor Statistics, "Databases, Tables & Calculators by Subject," n.d. As of November 26, 2014:
http://www.bls.gov/data/

U.S. Maritime Administration, *The Economic Importance of the U.S. Shipbuilding and Repairing Industry*, Washington, D.C., May 30, 2013.

Zycher, Benjamin, "Economic Effects of Reductions in Defense Outlays," *Policy Analysis*, No. 706, August 8, 2012.